An Introduction to Atmospheric Thermodynamics

An Introduction to Atmospheric Thermodynamics is a self-contained, concise but rigorous book introducing the reader to the basics of the subject.

Introductory chapters provide basic definitions and some useful mathematical and physical notes. Following a treatment of the fundamental laws of classical thermodynamics, the book describes topics relevant to atmospheric processes, including the properties of moist air and atmospheric stability. Thermodynamic diagrams are used as tools in the forecasting of storm development. In the final chapter the author introduces the problem of weather prediction and the relevance of thermodynamics.

The author has taught atmospheric thermodynamics at undergraduate level for 15 years and is a highly respected researcher in his field. This book provides an ideal text for short undergraduate courses taken as part of an atmospheric science, meteorology, physics or natural science program.

ANASTASIOS TSONIS received his Ph.D. in meteorology from McGill University, Canada, in 1982. He is currently Professor of Atmospheric Sciences in the Department of Mathematical Sciences at the University of Wisconsin, Milwaukee. Author of over 80 peer-reviewed papers, his main research interests include nonlinear dynamical systems and their application in climate, climate variability, and predictability. He has written two other books, *Chaos: From Theory to Applications* (1992) and, with J. B. Elsner, *Singular Spectrum Analysis: A New Tool in Time Series Analysis* (1996).

We have explained that the causes of the elements are four: the hot, the cold, the dry, and the moist. In every case, heat and cold determine, conjoin, and change things. Thus, hot and cold we describe as active, for combining is a sort of activity. Things dry and moist, on the other hand, are the subjects of that determination. In virtue of their being acted upon, they are thus passive.

Aristotle, Meteorology, Book IV

An Introduction to Atmospheric Thermodynamics

Anastasios A. Tsonis

University of Wisconsin–Milwaukee

CAMBRIDGE
UNIVERSITY PRESS

PUBLISHED BY THE PRESS SYNDICATE OF THE UNIVERSITY OF CAMBRIDGE
The Pitt Building, Trumpington Street, Cambridge, United Kingdom

CAMBRIDGE UNIVERSITY PRESS
The Edinburgh Building, Cambridge CB2 2RU, UK
40 West 20th Street, New York, NY 10011–4211, USA
477 Williamstown Road, Port Melbourne, VIC 3207, Australia
Ruiz de Alarcón 13, 28014 Madrid, Spain
Dock House, The Waterfront, Cape Town 8001, South Africa

http://www.cambridge.org

First published 2002

Printed in the United Kingdom at the University Press, Cambridge

Typeface Computer Modern 10/12pt.   *System* LaTeX 2$_\varepsilon$ [UPH]

A catalogue record for this book is available from the British Library

Library of Congress Cataloguing in Publication data

Tsonis, Anastasios A.
An introduction to atmospheric thermodynamics/Anastasios A. Tsonis
    p.   cm.
Includes bibliographical references and index.
ISBN 0 521 79263 0 – ISBN 0 521 79676 8 (pb.)
1. Atmospheric thermodynamics. I. Title.
QC880.4.T5 T76 2002
551.5′22–dc21   2001035622   CIP

ISBN 0 521 79263 0 hardback
ISBN 0 521 79676 8 paperback

CONTENTS

Preface   ix

1   Basic definitions   1

2   Some useful mathematical and physical topics   5
    2.1   Exact differentials   5
    2.2   Kinetic theory of heat   7

3   Early experiments and laws   11
    3.1   The first law of Gay-Lussac   11
    3.2   The second law of Gay-Lussac   12
    3.3   Absolute temperature   13
    3.4   Another form of the Gay-Lussac laws 13
    3.5   Boyle's law   14
    3.6   Avogadro's hypothesis   14
    3.7   The ideal gas law   15
    3.8   A little discussion on the ideal gas law 16
    3.9   Mixture of gases – Dalton's law   17
           Examples   19
           Problems   22

4   The first law of thermodynamics   23
    4.1   Work   23
    4.2   Definition of energy   25
    4.3   Equivalence between heat and work done 27
    4.4   Thermal capacities   28
    4.5   More on the relation between U and T (Joule's law)   30
    4.6   Consequences of the first law   33
           Examples   40
           Problems   46

5 The second law of thermodynamics   49
  5.1 The Carnot cycle   49
  5.2 Lessons learned from the Carnot cycle   52
  5.3 More on entropy   56
  5.4 Special forms of the second law   58
  5.5 Combining the first and second laws   60
  5.6 Some consequences of the second law   61
      Examples   67
      Problems   70

6 Water and its transformations   73
  6.1 Thermodynamic properties of water   74
  6.2 Equilibrium phase transformations – latent heat 77
  6.3 The Clausius–Clapeyron (C–C) equation   79
  6.4 Approximations and consequences of the C–C
      equation   81
      Examples   86
      Problems   89

7 Moist air   93
  7.1 Measures and description of moist air   93
    7.1.1 Humidity variables   93
    7.1.2 Mean molecular weight of moist air and
        other quantities   95
  7.2 Processes in the atmosphere   97
    7.2.1 Isobaric cooling – dew and frost temper-
        atures   97
    7.2.2 Adiabatic isobaric processes – wet-bulb
        temperatures   100
    7.2.3 Adiabatic expansion (or compression) of
        unsaturated moist air   103
    7.2.4 Reaching saturation by adiabatic ascent 104
    7.2.5 Saturated ascent   109
    7.2.6 A few more temperatures   113
    7.2.7 Saturated adiabatic lapse rate   114
  7.3 Other processes of interest   116
    7.3.1 Adiabatic isobaric mixing   116
    7.3.2 Vertical mixing   118
      Examples   120
      Problems   125

8 Vertical stability in the atmosphere   129
  8.1 The equation of motion for a parcel   129
  8.2 Stability analysis and conditions   131
  8.3 Other factors affecting stability   136
      Examples   136
      Problems   139

9   Thermodynamic diagrams   143
   9.1   Conditions for area-equivalent transformations   143
   9.2   Examples of thermodynamic diagrams   145
      9.2.1 The tephigram   145
      9.2.2 The emagram   147
      9.2.3 The skew emagram (skew $(T-\ln p)$
            diagram)   148
   9.3   Graphical representation of thermodynamic
       variables in a diagram   151
      9.3.1 Using diagrams in forecasting   152
      Examples   154
      Problems   156

10 Beyond this book   159
   10.1 Basic predictive equations in the atmosphere   159
   10.2 Moisture   161

References   163

Appendix   165
   Table A1   165
   Table A2   166
   Table A3   167
   The skew $(T-\ln p)$ diagram   168

Index   169

PREFACE

This book is intended for a semester undergraduate course in atmospheric thermodynamics. Writing it has been in my mind for a while. The main reason for wanting to write a book like this was that, simply, no such text in atmospheric thermodynamics exists. Do not get me wrong here. Excellent books treating the subject do exist and I have been positively influenced and guided by them in writing this one. However, in the past, atmospheric thermodynamics was either treated at graduate level or at undergraduate level in a partial way (using part of a general book in atmospheric physics) or too fully (thus making it difficult to fit it into a semester course). Starting from this point, my idea was to write a self-contained, short, but rigorous book that provides the basics in atmospheric thermodynamics and prepares undergraduates for the next level. Since atmospheric thermodynamics is established material, the originality of this book lies in its concise style and, I hope, in the effectiveness with which the material is presented. The first two chapters provide basic definitions and some useful mathematical and physical notes that we employ throughout the book. The next three chapters deal with more or less classical thermodynamical issues such as basic gas laws and the first and second laws of thermodynamics. In chapter six we introduce the thermodynamics of water, and in chapter seven we discuss in detail the properties of moist air and its role in atmospheric processes. In chapter eight we discuss atmospheric stability, and in chapter nine we introduce thermodynamic diagrams as tools to visualize thermodynamic processes in the atmosphere and to forecast storm development. Chapter ten serves as an epilogue and briefly discusses how thermodynamics blends into the weather prediction problem. At the end of each chapter solved examples are supplied. These examples were chosen to complement the theory and provide some direction for the unsolved problems. Instructors using this book may obtain the solutions to the unsolved problems directly from me.

Finally, I would like to extend my sincere thanks to Dr Vince Larson for critically proofreading this book, to Ms Gail Boviall for typing it and to Ms Donna Schenstrom for drafting the figures.

Anastasios A. Tsonis
Milwaukee

CHAPTER ONE

Basic definitions

- *Thermodynamics* is defined as the study of *equilibrium states* of a *system* which has been subjected to some *energy transformation*. More specifically, thermodynamics is concerned with transformations of heat into mechanical work and of mechanical work into heat.

- A *system* is a specific sample of matter. In the atmosphere a parcel of air is a system. A system is called *open* when it exchanges matter with its surroundings. In the atmosphere all systems are more or less open. A *closed* system is a system that does not exchange matter with its surroundings. In this case, the system is always composed of the same point-masses (a point-mass refers to a very small object, for example a molecule). Obviously, the mathematical treatment of closed systems is not as involved as the one for open systems, which are extremely hard to handle. Because of that, in atmospheric thermodynamics we assume that most systems are closed. This assumption is justified when the interactions associated with open systems can be neglected. This is approximately true in the following cases. (a) The system is large enough to ignore mixing with its surroundings at the boundaries. For example, a large cumulonimbus cloud may be considered as a closed system but a small cumulus may not. (b) The system is part of a larger homogeneous system. In this case mixing does not significantly change its composition. A system is called *isolated* when it exchanges neither matter nor energy with its surroundings.

- The *state* of a system (in classical mechanics) is completely specified at a given time if the position and velocity of each point-mass is known. Thus, in a three-dimensional world, for a system of N point-masses, $6N$ variables need to be known at any time. When N is very large (like in any parcel of air) this dynamical definition of state is not practical. As such, in thermodynamics we are dealing with the average properties of the system.

If the system is a homogeneous fluid consisting of just one component, then its thermodynamic state can be defined by its geometry, by its temperature, T, and pressure, p. The geometry of a system is defined by its volume, V, and its shape. However, most thermodynamic properties do not depend on shape. As such, volume is the only variable needed to characterize geometry. Since p, V, and T determine the state of the system, they must be connected. Their functional relationship $f(p, V, T) = 0$ is called the equation of state. Accordingly, any one of these variables can be expressed as a function of the other two. It follows that the state of a one-component homogeneous system can be completely defined by any two of the three state variables. This provides an easy way to visualize the evolution of such a system by simply plotting V against p in a rectangular coordinate system. In such a system, states of equal temperature define an isotherm. Any other thermodynamic variables that depend on the state defined by the two independent state variables are called state functions. State functions are thus dependent variables and state variables are independent variables; the two do not differ in other respects. That is why in the literature there is hardly any distinction between state variables and state functions. State variables and functions have the property that their changes depend only on the initial and final states, not on the particular way by which the change happened. If the system is composed of a homogeneous mixture of several components, then in order to define the state of the system we need, in addition to p, V, T, the concentrations of the different components. If the system is non-homogeneous we must divide it into a number of homogeneous parts. In this case, p, V, and T of a given homogeneous part are connected via an equation of state.

For a closed system, the chemical composition and its mass describe the system itself. Its volume, pressure, and temperature describe the state of system. Properties of the system are referred to as extensive if they depend on the size of the system and as intensive if they are independent of the size of the system. An extensive variable can be converted into an intensive one by dividing by the mass. In the literature it is common to use capital letters to describe quantities that depend on mass (work, W, entropy, S) and lower case letters to describe intensive variables (specific work, w, specific heat q). The mass, m, and temperature, T, will be exceptions to this rule.

- An *equilibrium state* is defined as a state in which the system's properties, so long as the external conditions (surroundings) remain unchanged, do not change in time. For example, a gas enclosed in a container of constant volume is in equilibrium if its pressure is constant throughout and its temperature is equal to that of the surroundings. An equilibrium state can be stable, unstable, or metastable. It is stable when small variations about the equilibrium state do not take the system away from the equilibrium state, and it is unstable if they do. An equilibrium state is called metastable if the system is stable with respect to small variations in certain properties and unstable with respect to small changes in other properties.

- A *transformation* takes a system from an initial state i to a final state f. In a (p, V) diagram such a transformation will be represented by a curve I connecting i and f. We will denote this as $i \xrightarrow{I} f$. A transformation can be reversible or irreversible. Formally, a reversible transformation is one in which the successive states (those between i and f) differ by infinitesimals from equilibrium states. Accordingly, a reversible transformation can only connect those i and f states which are equilibrium states. It follows that a reversible process is one which can be reversed anywhere along its path in such a way that both the system and its surroundings return to their initial states. In practice a reversible transformation is realized only when the external conditions change very slowly so that the system has time to adjust to the new conditions. For example, assume that our system is a gas enclosed in a container with a movable piston. As long as the piston moves from i to f very slowly the system adjusts and all intermediate states are equilibrium states. If the piston does not move slowly, then currents will be created in the expanding gas and the intermediate states will not be equilibrium states. From this example, it follows that turbulent mixing in the atmosphere is a source of irreversibility. If a system goes from i to f reversibly, then it could go from f to i again reversibly if the same steps were followed backwards. If the same steps cannot be followed exactly, then this transformation is represented by another curve I' in the (p, V) diagram (*i.e.* $f \xrightarrow{I'} i$) and may or may not be reversible. In other words the system may return to its initial state but the surroundings may not. Any transformation $i \longrightarrow f \longrightarrow i$ is called a cyclic transformation. Given the discussion above we can have cyclic transformations which are reversible or irreversible (figure 1.1). A transformation $i \xrightarrow{I} f$ is called isothermal if I is an isotherm, isochoric if I is a constant volume line, isobaric if I is a constant pressure line, and adiabatic if during the transformation the system does not exchange heat with its surroundings (environment). Note and keep in mind for later that adiabatic transformations are not isothermal.

- *Energy* is something that can be defined formally (we have to

Figure 1.1
(a) A reversible cyclic
process; (b) an irreversible
cyclic process.

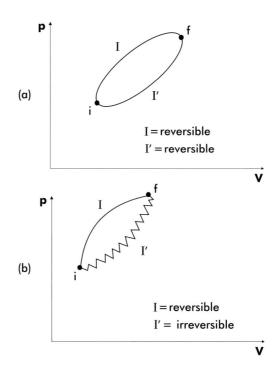

wait a bit for this) but its concept is not easily understood by
defining it. We all feel we understand what is meant by energy,
but if we verbally attempt to explain what energy is we will get
upset with ourselves. At this point, let us just recall that for
a point-mass with a mass m_{p} moving with speed v in a uniform gravitational field g, Newton's second law takes the form
$d(K + P)/dt = 0$ where $K = m_{\mathrm{p}}v^2/2$, $P = m_{\mathrm{p}}gz$, t is the
time, and z is the height. K is called the kinetic energy and P is
called the potential energy. The total energy of the point-mass
$E = K + P$ is, therefore, conserved. If we consider a system of
N interacting point-masses that may be subjected to external
forces (other than gravity), then the total energy of the system
is the sum of the kinetic energy about the centre of gravity of
all point-masses (internal kinetic energy), the kinetic energy of
the centre of gravity, the potential energy due to interactions
between the point-masses (internal potential energy), and the
potential energy due to external forces. The sum of the internal
kinetic and internal potential energy is called the internal energy
of the system, U. A system is called conservative if $dU/dt = 0$
and dissipative otherwise. For systems considered here we are
interested in the internal energies only.

CHAPTER TWO

Some useful mathematical and physical topics

2.1 Exact differentials

If z is a function of the variables x and y, then by definition

$$dz = \left(\frac{\partial z}{\partial x}\right)_y dx + \left(\frac{\partial z}{\partial y}\right)_x dy \tag{2.1}$$

where dz is an exact differential. Now let us assume that a quantity δz can be expressed according to the following differential relationship

$$\delta z = M dx + N dy \tag{2.2}$$

where x and y are independent variables and M and N are functions of x and y. If we integrate equation (2.2) we have that

$$\int \delta z = \int M dx + \int N dy.$$

Since M and N are functions of x and y, the above integration cannot be done unless a functional relationship $f(x, y) = 0$ between x and y is chosen. This relationship defines a path in the (x, y) domain along which the integration will be performed. This is called a *line integral* and its result depends entirely on the prescribed path in the (x, y) domain. If it is that

$$M = \frac{\partial z}{\partial x}, \ N = \frac{\partial z}{\partial y} \tag{2.3}$$

then equation (2.2) becomes

$$\delta z = \frac{\partial z}{\partial x} dx + \frac{\partial z}{\partial y} dy.$$

The right-hand side of the above equation is the exact or total differential dz. In this case δz is an exact differential. If we now integrate δz from some initial state i to a final state f we obtain

$$\int_i^f \delta z = \int_i^f dz = z(x_f, y_f) - z(x_i, y_i). \tag{2.4}$$

Clearly, if δz is an exact differential its net change along a path $i \longrightarrow f$ depends only on points i and f and not on the particular path from i to f. We say that in this case z is a point function. All three state variables are exact differentials (i.e. $\delta p = dp$, $\delta T = dT$, $\delta V = dV$). It follows that all quantities that are a function of any two state variables will be exact differentials.

If the final and initial states coincide (i.e. we go back to the initial state via a cyclic process), then from equation (2.4) we have that

$$\oint \delta z = 0. \tag{2.5}$$

The above alternative condition indicates that δz is an exact differential if its integral along any closed path is zero. At this point we have to clarify a point so that we do not get confused later. When we deal with pure mathematical functions our ability to evaluate $\oint \delta z$ does not depend on the direction of the closed path, i.e. whether we go from i back to i via $i \xrightarrow{I} f \xrightarrow{I'} i$ or via $i \xrightarrow{I'} f \xrightarrow{I} i$ (see figure 1.1). When we deal with natural systems we must view the condition $\oint \delta z = 0$ in relation to reversible and irreversible processes. If, somehow, it is possible to go from i back to i via $i \xrightarrow{I} f \xrightarrow{I'} i$ but impossible via $i \xrightarrow{I'} f \xrightarrow{I} i$ (for example, when I' is an irreversible transformation), then computation of δz depends on the direction and as such it is not unique. Therefore, for physical systems, the condition $\oint \delta z = 0$ when δz is an exact differential applies only to reversible processes.

Note that since

$$\frac{\partial}{\partial y}\frac{\partial z}{\partial x} = \frac{\partial^2 z}{\partial y \partial x} = \frac{\partial^2 z}{\partial x \partial y} = \frac{\partial}{\partial x}\frac{\partial z}{\partial y}$$

it follows that an equivalent condition for δz to be an exact differential is that

$$\frac{\partial M}{\partial y} = \frac{\partial N}{\partial x}. \tag{2.6}$$

Equations (2.3)–(2.6) are equivalent conditions that define z as a point function and subsequently δz as an exact differential. If a thermodynamic quantity is not a point function or an exact differential then its change along a path depends on the path. Moreover, its change along a closed path is not zero. Such quantities are path functions. For a path function the thermodynamic processes must be specified completely in order to define the quantity. For the rest of this book an exact differential will be denoted by dz whereas a non-exact differential will be denoted by δz. Finally, note that if δz

is not an exact differential and if only two variables are involved, a factor λ (called the integration factor) may exist such that $\lambda \delta z$ is an exact differential.

2.2 Kinetic theory of heat

Let us consider a system at a temperature T, consisting of N point-masses (molecules). According to the kinetic theory of heat, these molecules move randomly at all directions traversing rectilinear lines. This motion is called Brownian motion. Because of the complete randomness of this motion, the internal energies of the point-masses not only are not equal to each other, but they change in time. If, however, we calculate the mean internal energy, we will find that it remains constant in time. The kinetic theory of heat accepts that the mean internal energy of *each* point-mass, $\overline{U}$, is proportional to the absolute temperature of the system (a formal definition of absolute temperature will come later; for now let us denote it by T),

$$\overline{U} = \text{constant} \times T \qquad (2.7)$$

Let us for a minute assume that $N = 1$. Then the point has only three degrees of freedom which here are called thermodynamic degrees of freedom and are equal to the number of independent variables needed to completely define the energy of the point (unlike the degrees of freedom in Hamiltonian dynamics which are defined as the least number of independent variables that completely define the position of the point in state space). The velocity v of the point can be written as

$$v^2 = v_x^2 + v_y^2 + v_z^2.$$

Because we only assumed one point, then the total internal energy is equal to its kinetic energy. Thus,

$$U = \frac{m_{\mathrm{p}} v^2}{2}$$

or

$$U_x = \frac{m_{\mathrm{p}} v_x^2}{2}, \ U_y = \frac{m_{\mathrm{p}} v_y^2}{2}, \ U_z = \frac{m_{\mathrm{p}} v_z^2}{2}$$

where to each component corresponds one degree of freedom. According to the equal distribution of energy theorem, the mean kinetic energy of the point, $\overline{U}$, is distributed equally to the three degrees of freedom i.e. $\overline{U}_x = \overline{U}_y = \overline{U}_z$. Accordingly, from equation (2.7) we can write that

$$\overline{U}_i = AT, \quad i = x, y, z$$

where the constant A is a universal constant (i.e. it does not depend on the degrees of freedom or the type of the gas). We denote this constant as $k/2$ where k is Boltzmann's constant ($k = 1.38 \times 10^{-23}$ J K^{-1}) (for a review of units, see Table A1 in the Appendix). Therefore, the mean kinetic energy of a point with three degrees of freedom is equal to

$$\overline{U} = \frac{3kT}{2} \tag{2.8}$$

or

$$\frac{\overline{m_\mathrm{p} v^2}}{2} = \frac{3kT}{2}.$$

The theorem of equal energy distribution can be extended to N points. In this case, the degrees of freedom are $3N$ and

$$\sum_{i=1}^{N} \frac{\overline{m_\mathrm{p} v_i^2}}{2} = \frac{3NkT}{2}$$

or

$$\frac{1}{N} \sum_{i=1}^{N} \frac{\overline{m_\mathrm{p} v_i^2}}{2} = \frac{3}{2} kT$$

or

$$\left\langle \frac{\overline{m_\mathrm{p} v^2}}{2} \right\rangle = \frac{3}{2} kT \tag{2.9}$$

where $\langle \frac{\overline{m_\mathrm{p} v^2}}{2} \rangle$ is the average kinetic energy of *all* N points. Note that the above is true only if the points are considered as monatomic. If they are not, extra degrees of freedom are present that correspond to other motions such as rotation about the center of gravity, oscillation about the equilibrium positions, etc.

The kinetic theory of heat has found many applications in the kinetic theory of *ideal gases*. An ideal gas is one for which the following apply:

(a) the molecules move randomly in all directions and in such a way that the same number of molecules move in any direction;

(b) during the motion the molecules do not exert forces except when they collide with each other or with the walls of the container. As such the motion of each molecule between two collisions is linear and of uniform speed;

(c) the collisions between molecules are considered elastic. This is necessary because otherwise with each collision the kinetic energy of the molecules will be reduced thereby resulting in a temperature decrease. Also, a collision obeys the law of specular reflection (the angle of incidence equals the angle of reflection);

(d) the sum of the volumes of the molecules is negligible compared with the volume of the container.

Figure 2.1
A molecule of mass m_p moving with a velocity v and hitting a surface S. If this collision is assumed elastic and specular, then the change in momentum is $2m_\mathrm{p}v$.

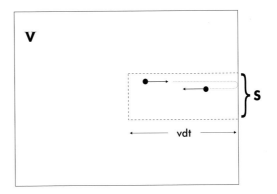

Now let us consider a molecule of mass m_p whose velocity is v and which is moving in a direction perpendicular to a wall (figure 2.1). The molecule has a momentum $P = m_\mathrm{p}v$. Since we accept that the collision is elastic and specular, the magnitude of the momentum after the collision is $-m_\mathrm{p}v$. Thus the total change in momentum is $m_\mathrm{p}v - (-m_\mathrm{p}v) = 2m_\mathrm{p}v$. According to Newton's second law, $F = dP/dt$. If we consider N molecules occupying a volume V we can calculate the change in momentum dP of all molecules in the time interval dt by multiplying the change in momentum of one molecule $(2m_\mathrm{p}v)$ by the number of molecules dN that hit a given area S on the wall, i.e.

$$dP = 2m_\mathrm{p}vdN \tag{2.10}$$

Note that here we have assumed that all molecules have the same speed. The number dN of molecules hitting area S during dt is equal to the number of molecules which move to the right and which are included in a box with base S and length vdt. Since the motion is completely random, we can assume that $\frac{1}{6}$ of the molecules will be moving to the right, $\frac{1}{6}$ will be moving to the left, and $\frac{4}{6}$ will be moving along the directions of the other two coordinates. Since the volume of the box is $Svdt$ and the number of molecules per unit volume is N/V then the number of molecules inside the box is

$$\frac{N}{V}Svdt.$$

Accordingly, the number of molecules moving to the right and colliding with S is

$$dN = \frac{N}{6V}Svdt \tag{2.11}$$

From equations (2.10) and (2.11) it follows that

$$dP = 2m_\mathrm{p}v^2\frac{N}{6V}Sdt.$$

Recalling the definition of pressure, p (pressure = force/area), and Newton's second law we obtain

$$p = \frac{1}{3}\frac{N}{V}m_\mathrm{p}v^2.$$

The above formula resulted by assuming that all molecules move with the same speed. This is not true, and because of that, $m_\mathrm{p}v^2$ in the above equation should be replaced by the average of all points, $\langle\overline{m_\mathrm{p}v^2}\rangle$. We thus arrive at

$$p = \frac{1}{3}\frac{N}{V}\langle\overline{m_\mathrm{p}v^2}\rangle. \tag{2.12}$$

It can easily be shown that by combining equations (2.9) and (2.12), we can derive an equation that includes all three state variables:

$$pV = NkT. \tag{2.13}$$

Equation (2.13) provides the functional relationship of the equation of state $f(p, V, T)$ and it is called the ideal gas law. More details follow in the next chapter.

CHAPTER THREE

Early experiments and laws

At the end of Chapter 2 we derived theoretically the equation of state or the ideal gas law. This law was first derived experimentally. The relevant experiments provide many interesting insights about the properties of ideal gases and confirm the theory. As such a little discussion is necessary.

3.1 The first law of Gay-Lussac

Through experiments Gay-Lussac was able to show that, when pressure is constant, the increase in volume of an ideal gas, dV, is proportional to the volume V_0 that it has at a temperature (measured in the Celsius scale) of $\theta = 0\,°C$ and proportional to the temperature increase, $d\theta$:

$$dV = \alpha V_0 d\theta. \qquad (3.1)$$

The coefficient α is called the volume coefficient of thermal expansion at a constant pressure and it has the value of $1/273\,\mathrm{deg}^{-1}$ for all gases. The physical meaning of α can be understood if we solve equation (3.1) for α:

$$\alpha = \frac{1}{d\theta}\frac{dV}{V_0}.$$

From the above equation it follows that if we increase the temperature of an ideal gas by $1\,°C$, while we keep the pressure constant, the volume will increase by $1/273$ of the volume the gas occupies at $0\,°C$. By integrating equation (3.1) we obtain the relationship between V and θ:

$$\int_{V_0}^{V} dV = \int_{0}^{\theta} \alpha V_0 d\theta$$

or

$$V - V_0 = \alpha V_0 \theta$$

Figure 3.1
Graphical representation
of the first law of
Gay-Lussac.

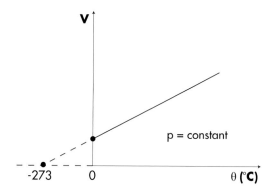

or

$$V = V_0(1 + \alpha\theta). \tag{3.2}$$

This is a linear relationship and its graph is shown in figure 3.1.

3.2 The second law of Gay-Lussac

Again through experimentation Gay-Lussac was able to show that for a constant volume the increase in pressure of an ideal gas, dp, is proportional to the pressure p_0 that it has at a temperature of $0\,^\circ\text{C}$ and proportional to the increase in temperature $d\theta$:

$$dp = \beta p_0 d\theta.$$

The coefficient β is the pressure coefficient of thermal expansion at a constant volume and has the value $1/273\,\text{deg}^{-1}$ for all gases:

$$\beta = \frac{1}{d\theta}\frac{dp}{p_0}.$$

The above formula indicates that an increase in temperature by $1\,^\circ\text{C}$ (while V is constant) results in an increase in pressure by $1/273$ of the pressure the gas had at $0\,^\circ\text{C}$.

As in the first law again it follows that (see figure 3.2)

$$p = p_0(1 + \beta\theta). \tag{3.3}$$

Application

The second law of Gay-Lussac can easily explain why in the winter heating a house at a much greater temperature than the outside temperature does not increase the pressure enough to break the windows. A difference of $20\,^\circ\text{C}$ between the inside and the outside air increases the pressure inside the house by 7.3%. Glass can withstand such pressure changes easily.

Figure 3.2
Graphical representation
of the second law of
Gay-Lussac.

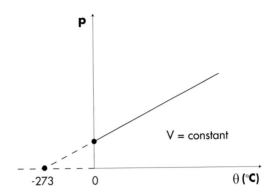

3.3 Absolute temperature

From equation (3.2) it follows that if we extrapolate to $\theta = -273\,°\text{C}$, then $V = 0$. This means that if we were able to cool an ideal gas to $-273\,°\text{C}$, while keeping the pressure constant, the volume would become zero. Similarly, from equation (3.3) it follows that if we were able to cool an ideal gas to the temperature of $-273\,°\text{C}$, while keeping the volume constant, the pressure would become zero. This temperature of $-273\,°\text{C}$ we call absolute zero. Up to now for the measurement of temperature the Celsius scale has been used, starting from the temperature "zero Celsius". If we extend the Celsius scale at temperatures below zero and as the beginning of the scale we consider the absolute zero (i.e. $-273\,°\text{C}$), then the temperature measured from the absolute zero is called absolute temperature, T. This new scale is called the Kelvin scale ($T = 273 + \theta$).

3.4 Another form of the Gay-Lussac laws

Using the absolute temperature we can present the Gay-Lussac laws as follows. From figure (3.3) it follows from the similarity of triangles ABC and $AB'C'$ that his first law can be expressed as

$$\text{for } p = \text{ constant}, \quad \frac{V}{V'} = \frac{T}{T'}. \tag{3.4}$$

Similarly (figure 3.4), his second law can be expressed as

$$\text{for } V = \text{ constant}, \quad \frac{p}{p'} = \frac{T}{T'}. \tag{3.5}$$

Verbally stated, the volumes of an ideal gas under constant pressure and the pressures under constant volume are proportional to the absolute temperatures.

Figure 3.3
Expressing Gay-Lussac's
first law for two states C
and C'.

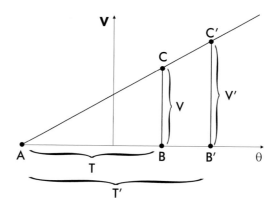

Figure 3.4
Expressing Gay-Lussac's
second law for two states
C and C'.

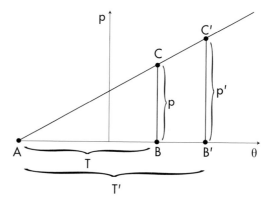

3.5 Boyle's law

While Gay-Lussac's laws provide the change in volume or pressure
as a function of temperature, Boyle's law provides the change in
pressure as the volume varies at a constant temperature. The law
is expressed as follows:

$$pV = p'V'. \tag{3.6}$$

3.6 Avogadro's hypothesis

From chemistry it is known that during chemical reactions between
various gases there is a simple *proportion* between the volumes
of the reactant gases and volumes of the products. This is called
the law of volumes. For example, under the condition that the
comparison is made at the same temperature and pressure, it is
found that in the reaction $2CO + O_2 \rightarrow 2CO_2$ two volumes of CO

react with one volume of O_2 to produce two volumes of CO_2. The law of volumes is explained easily if we accept that *equal volumes of gases under constant temperature and pressure contain the same number of molecules independently of the gas*. This is Avogadro's hypothesis. In the above example, and according to this hypothesis, if the considered amount of O_2 contains N molecules then since the CO and the CO_2 each occupy double the volume they should each contain $2N$ molecules. Recalling that in a mole of any gas there are $N = 6.022 \times 10^{23}$ molecules, then Avogadro's hypothesis can be stated simply as: *A mole of any gas at constant temperature and pressure occupies the same volume*. This volume for the standard state $T_0 = 0\,°C$, $p_0 = 1$ atmosphere has the value (see equation 2.13)

$$V_{T_0,p_0} = 22.4 \quad \text{liters mol}^{-1}$$
$$= 22\,400 \text{ cm}^3 \text{ mol}^{-1}.$$

3.7 The ideal gas law

Let us consider an ideal gas at a state p, V, T which is heated under constant volume to a state p_1, V, T'. Then, according to Gay-Lussac's 2nd law

$$p_1 = p\frac{T'}{T}.$$

If subsequently we keep the temperature constant and increase the volume to V' the gas goes to a state $p'V'T'$. Then, according to Boyle's law,

$$p'V' = p_1 V.$$

By combining the above two equations we obtain

$$p'V' = p\frac{T'}{T}V$$

or

$$\frac{pV}{T} = \frac{p'V'}{T'}. \tag{3.7}$$

This is called the Boyle–Gay-Lussac law and it indicates that if we change the pressure, the volume, and the temperature the value of PV/T remains constant. Thus, equation (3.7) can be written as

$$pV = AT \tag{3.8}$$

where A is, according to equation (3.7), a constant depending on the type and mass of the gas. The dependence on the mass follows from the fact that if we consider twice the mass and keep the pressure and the temperature constant, then the volume will double (Avogadro's hypothesis). Therefore, the value of pV/T (i.e. A) will double. Accordingly, we can write $A = mR$ where m is the mass of the gas and R is a constant independent of the mass but dependent on the gas. This constant is called the specific gas constant. We can then write equation (3.8) as

$$pV = mRT. \qquad (3.9)$$

Since $m = Nm_{\mathrm{p}}$ (where N is the total number of molecules each having mass m_{p}), equation (3.9) is identical to equation (2.13) with $R = k/m_{\mathrm{p}}$. Equation (3.9) is thus the equation of state or ideal gas law which is now derived experimentally. Equation (3.9) can be written as $p = \rho RT$ where $\rho = m/V$ is the mass density. By defining the specific volume $a = 1/\rho$ the ideal gas law takes the form

$$pa = RT.$$

If M denotes the molecular weight of the gas then the number of moles is $n = m/M$. It follows that

$$pV = nMRT$$

or

$$pV = nR^*T \qquad (3.10)$$

or

$$pa = \frac{R^*}{M}T$$

where $R^* = MR$. Considering that $R = k/m_{\mathrm{p}}$ it follows that $R^* = Mk/m_{\mathrm{p}} = mk/nm_{\mathrm{p}} = Nk/n$. Now recall that the number of molecules in one mole, N_A, is equal to 6.022×10^{23} (Avogadro's number). Then, $N = nN_A$. In this case $R^* = N_A k$, which is the product of two constants. This new constant is called the universal gas constant and its value is 8.3143 J K^{-1} mol^{-1}.

3.8 A little discussion on the ideal gas law

- Because the ideal gas law relates three variables (rather than two), we have to be careful when we interpret changes in one of the variables. For example, temperature increases when pressure increases but only if the volume (density) increases (decreases) or remains constant or decreases (increases) by a smaller amount compared with pressure. Let us consider the ideal gas law $pV = mRT$ or $p = \rho RT$ and differentiate it:

$$pdV + Vdp = mRdT$$

$$dp = RTd\rho + R\rho dT.$$

We can easily see that if $dp > 0$ and $dV \geq 0$ then $dT > 0$. Also, if $dp > 0$ and $dV < 0$ and $|Vdp| > |pdV|$ then $dT > 0$. And so on!

It follows that cold air is denser than warm air only if pressure remains constant or if its change does not offset the temperature difference.

Similarly, if we consider the ideal gas law in the form

$$pa = \frac{R^*}{M}T$$

or

$$p = \rho\frac{R^*}{M}T$$

we can reason that moist air (in which some of the dry air of molecular weight 28 g mol^{-1} has been replaced by water vapor of molecular weight 18 g mol^{-1}) is less dense than dry air at more or less the same pressure and temperature. This explains some interesting statistics in the American game of baseball where more home runs occur when the weather is hot and humid. In a warmer and more moist environment where $p = $ constant, the density of air is smaller. Therefore, the ball has less resistance. For a path of 400 feet or so the effect can be significant, resulting in higher chances for a home run.

- Recall equations (2.9) and (3.2). According to equation (2.9) $T = 0$ K when $v = 0$. Because of this one might interpret the zero absolute temperature as the temperature where all motion in an ideal gas ceases. On the other hand, when equation (3.2) is extrapolated to the temperature at which $V = 0$ it results in $T_{\text{absolute}} = (-\frac{1}{\alpha})\,^{\circ}\text{C} = -273\,^{\circ}\text{C} = 0$ K. While this temperature corresponds to vanishing volumes, it is not necessarily the temperature at which motion ceases. If both $V = 0$ and $v = 0$ are true when $T = 0$ K, then pressure can be anything, but this does not invalidate the ideal gas law, both sides of which will equal zero. To avoid confusion, we must keep in mind that these issues emerge when these equations are extrapolated to states where one cannot assume that a gas behaves as an ideal gas. At those states, equations (2.9) and (3.2) simply do not mean much.

3.9 Mixture of gases – Dalton's law

Consider a mixture of two gases occupying a volume V and consisting of N_1 molecules of gas 1 and N_2 molecules of gas 2. The total pressure on the walls will be the result of all the collisions, i.e. the collisions by the molecules of gas 1 and the collisions by the molecules of gas 2. Thus we can write equation (2.13) as

$$p = \frac{(N_1 + N_2)}{V}kT$$

or

$$p = \frac{N_1 kT}{V} + \frac{N_2 kT}{V}.$$

The first term on the right-hand side of the above equation is exactly the pressure that gas 1 would create if all its N_1 molecules occupied the volume V, i.e. the partial pressure of gas 1. The same is valid for the second term. We thus conclude that the total pressure is the sum of the partial pressures. This expresses Dalton's law which states that for a mixture of K components, each one of which obeys the ideal gas law, the total pressure, p, exerted by the mixture is equal to the sum of the partial pressures which would be exerted by each gas if it alone occupied the entire volume at the temperature of the mixture, T:

$$p = \sum_{i=1}^{K} p_i.$$

If the volume of the mixture is V and the mass and molecular weight of the ith constituent are m_i and M_i respectively, then for each constituent

$$p_i = \frac{R^*}{M_i} \frac{m_i}{V} T.$$

Applying Dalton's law, we then can write that for a mixture

$$p = \sum_{i=1}^{K} \frac{R^* T}{V} \frac{m_i}{M_i}$$

or

$$p = \frac{R^* T}{V} \sum_{i=1}^{K} \frac{m_i}{M_i}.$$

Since the total mass of the mixture $m = \sum_{i=1}^{K} m_i$, the above equation becomes

$$p = \frac{R^* T m}{V} \frac{\sum_{i=1}^{K} \frac{m_i}{M_i}}{\sum_{i=1}^{K} m_i}$$

or

$$p = \frac{R^* T}{a} \frac{\sum_{i=1}^{K} \frac{m_i}{M_i}}{\sum_{i=1}^{K} m_i} \tag{3.11}$$

In order for the mixture to obey the ideal gas law, it must satisfy the equation

$$p = \frac{RT}{a} = \frac{R^* T}{a \overline{M}} \tag{3.12}$$

where $\overline{M}$ is the mean molecular weight of the mixture. By comparing equations (3.11) and (3.12) we see that this would be possible if

$$\overline{M} = \frac{\sum_{i=1}^{K} m_i}{\sum_{i=1}^{K} \frac{m_i}{M_i}}.$$ (3.13)

This provides the proper way to compute the mean molecular weight for a mixture and indicates that the mixture follows the ideal gas law as long as there is no condensation (if there is, then some m_i's may not remain constant).

Examples

(3.1)   Determine the mean molecular weight of dry air.

For our planet, the lowest 25 km of the atmosphere is made up almost entirely by nitrogen (N_2), oxygen (O_2), argon (A), and carbon dioxide (CO_2) (75.51%, 23.14%, 1.3%, and 0.05% by mass, respectively). Thus, for dry air the mean molecular weight is

$$\overline{M} = \frac{m_{N_2} + m_{O_2} + m_A + m_{CO_2}}{\frac{m_{N_2}}{M_{N_2}} + \frac{m_{O_2}}{M_{O_2}} + \frac{m_A}{M_A} + \frac{m_{CO_2}}{M_{CO_2}}}$$

or

$$\overline{M} = \frac{75.51 + 23.16 + 1.3 + 0.05}{\frac{75.51}{28.02} + \frac{23.14}{32.0} + \frac{1.3}{39.94} + \frac{0.05}{44.01}} \text{ g mol}^{-1}$$

or

$$\overline{M} = 28.97 \text{ g mol}^{-1}$$

or

$$\overline{M} = 0.02897 \text{ kg mol}^{-1}.$$

It follows that the specific gas constant for dry air is

$$R_d = R^*/\overline{M} = 287 \text{ J kg}^{-1}\text{K}^{-1}.$$

(3.2)   Determine the mean molecular weight of a mixture of dry air saturated with water vapor at 0 °C and 1 atmosphere pressure. The partial pressure of water vapor at 0 °C is 6.11 mbar.

The mixture consists of dry air and water vapor. If p_d is the pressure due to dry air and p_v the pressure due to water vapor, then the pressure of the mixture is $p = p_d + p_v$. If we recall that the number of moles $n = m/M$ then we can write equation (3.13) as

$$\overline{M} = \frac{\sum_{i=1}^{K} n_i M_i}{\sum_{i=1}^{K} n_i} = \frac{\sum_{i=1}^{K} n_i M_i}{n} = \sum_{i=1}^{K} \frac{n_i}{n} M_i$$ (3.14)

where n_i and n are the corresponding number of moles of the constituents of the mixture and the total number of moles in the mixture. Assuming that both dry air and water vapor are ideal gases we have that

$$pV = nR^*T \quad \text{for the mixture}$$
$$p_iV = n_iR^*T \quad \text{for each of the constituents.}$$

From these two equations it follows that

$$\frac{p}{p_i} = \frac{n}{n_i}.$$

Combining the above relationship and equation (3.14) yields

$$\overline{M} = \sum_{i=1}^{K} \frac{p_i}{p} M_i.$$

In our case $i = 1, 2$. Thus,

$$\overline{M} = \frac{p_1}{p} M_1 + \frac{p_2}{p} M_2$$

or changing $1 \rightarrow$ d (dry air) and $2 \rightarrow$ v (water vapor)

$$\overline{M} = \frac{p_\text{d}}{p} M_d + \frac{p_\text{v}}{p} M_\text{v}$$

or

$$\overline{M} = \frac{p - p_\text{v}}{p} M_d + \frac{p_\text{v}}{p} M_\text{v}$$

or

$$\overline{M} = 28.9 \text{ g mol}^{-1}.$$

The above estimated mean molecular weight is not very much different from the mean molecular weight of dry air. The difference becomes more apparent for higher temperatures. For example for $T = 35\,^\circ$C the partial pressure of water vapor is 57.6 mbar and $\overline{M} = 28.3$ g mol^{-1} .

(3.3) Two containers A and B of volumes $V_\text{A} = 800$ cm^3 and $V_\text{B} = 600$ cm^3, respectively, are connected with a tube that closes and opens by a hinge. The containers are filled with a gas under pressures of 1000 mbar and 800 mbar, respectively. If we open the connection what will the final pressure be in each of the containers? Assume that the temperature remains constant.

Once the connection is open, each gas will expand to fill the total volume $V = V_\text{A} + V_\text{B}$ thereby equalizing the difference in pressure between the two containers. Thus the final pressure in each container will be the same. Let us first assume that container B is empty. Since the gas in container

A expands at a constant temperature we have (Boyle's law)

$$p_f V_f = p_i V_i$$

where i and f stand for initial and final. It follows that $p_f = 571$ mbar.

Similarly if we assume that container A is empty the final pressure in both containers will be

$$p_f' = \frac{p_i V_i}{V_f} = 343 \text{ mbar.}$$

Since $p_f(p_f')$ is the pressure the gas in A(B) would exert if it occupied the total volume $V_A + V_B, p_f$ and p_f' can be considered as partial pressures of two gases. Then from Dalton's law it follows that the final pressure in each container should be 914 mbar.

(3.4)   If Boyle had observed $\sqrt{p}V =$ constant, what would the equation of state for an ideal gas be? In this case calculate the temperature of a sample of nitrogen which has a pressure of 800 mbar and a specific volume of 1200 cm^3 g^{-1}. How does this temperature differ from that estimated when the correct law $pV =$ constant is observed? Can you explain the difference?

If we follow the procedure to arrive at equation (3.7) but with the new Boyle's law, we have that

$$p_1 = p\frac{T'}{T}$$

$$\sqrt{p'}V' = \sqrt{p_1}V.$$

By combining the above equations we have that

$$\sqrt{p'}V' = \sqrt{p}\sqrt{\frac{T'}{T}}V$$

or

$$\frac{\sqrt{p'}V'}{\sqrt{T'}} = \frac{\sqrt{p}V}{\sqrt{T}} = \text{constant.}$$

As we know if we keep the temperature and pressure constant and double the mass, the volume will double. As such, here as well the ratio $\sqrt{p}V/\sqrt{T}$ will double. Then we can write that $\sqrt{p}V/\sqrt{T} = mR$ where m is the total mass and R is the specific gas constant. Thus, the ideal gas law in this case will be

$$\sqrt{p}a = \frac{R^*}{M}\sqrt{T}.$$

Solving for T we have that

$$\sqrt{T} = 1.143$$

or
$$T = 1.07\,\text{K}.$$

Under normal circumstances (i.e. when $pa = (R^*/M)T$), we find that $T = 323.3\,\text{K}$ which makes more sense. The tremendous difference is due to the fact that the square root in $\sqrt{p}a = \text{constant}$ introduces two corrections to the ideal gas law (both pressure and temperature appear under a square root) having the net result of significantly reducing T.

Problems

(3.1) What is the mass of dry air occupying a room of dimensions $3 \times 5 \times 4$ m at $p = 1$ atmosphere and $T = 20\,^\circ\text{C}$? (72.3 kg)

(3.2) Graph the relationship $V = f(T)_{p=\text{constant}}$ from absolute zero up to high temperatures, for two samples of the same gas which at $0\,^\circ\text{C}$ occupy volumes of 1000 and 2000 cm^3, respectively.

(3.3) In a 2-D coordinate system with axes the absolute temperature and volume, graph (a) an isobaric (constant pressure) change of one mole of an ideal gas for $p = 1$ atmosphere, and (b) the same when $p = 2$ atmospheres.

(3.4) Determine the molecular weight of the Venusian atmosphere assuming that it consists of 95% CO_2 and 5% N_2 by volume. What is the gas constant for 1 kg of such an atmosphere? (43.2 g mol^{-1}, 192.5 J kg^{-1} K^{-1})

(3.5) If $p = 1$ atmosphere and $T = 0\,^\circ\text{C}$ how many molecules are there in 1 cm^3 of dry air? (2.6884×10^{19} molecules)

(3.6) Given the two states p, V, T and p', V', T', define on a (p, V) diagram the state p_1, V, T' that was used to arrive at the ideal gas law (section 3.7).

(3.7) An ideal gas of p, V, T undergoes the following successive changes: (1) it is warmed under a constant pressure until its volume doubles, (2) it is warmed under constant volume until its pressure doubles, and (3) it expands isothermally until its pressure returns to p. Calculate in each case the values of p, V, T and plot all three changes in a (p, V) diagram.

CHAPTER FOUR

The first law of thermodynamics

4.1 Work

As we have already mentioned, in atmospheric thermodynamics we will be dealing with equilibrium states of air. If a system (parcel of air, for example), is at equilibrium with its environment no changes take place in either of them. We can imagine the "shape" of the parcel remaining unchanged in time. If the pressure of the surroundings changes, then the force associated with the pressure change will disturb the parcel thereby forcing it away from equilibrium. In order for the parcel to adjust to the pressure changes of the surroundings, the parcel will either contract or expand. If the parcel expands we say that the parcel performs work on the environment and if the parcel contracts we say that the environment performs work on the parcel. By definition, if the volume change is dV then the incremental work done, dW, is

$$dW = pdV.$$

Accordingly, when the system changes from an initial state i to a final state f the total work done, either by the system or on the system, is

$$W = \int_i^f pdV.$$

The above equation indicates that the work done is given by an area in a (p, V) diagram (figure 4.1) [or a (p, a) diagram, where a is the specific volume]. If $dV > 0$ (the system expands) it follows that $W > 0$ and if $dV < 0$ (system contracts) it follows that $W < 0$. Thus, positive work corresponds to work done by the system on the environment and negative work corresponds to work done to the system by the environment.

Now let us consider a situation where the system expands through a reversible transformation from i to f and then contracts

Figure 4.1
The shaded area gives the work done when a system changes from an initial state i to a final state f.

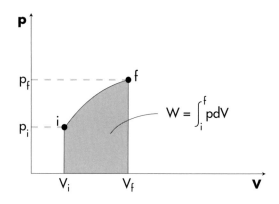

Figure 4.2
The shaded area gives the work done during a cyclic reversible transformation.

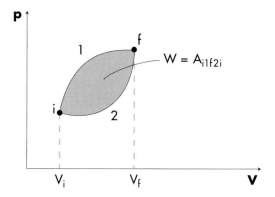

from f to i along exactly the same path in the (p, V) diagram. Then, the total work done will be

$$W = \int_i^f pdV + \int_f^i pdV = \int_i^f pdV - \int_i^f pdV = 0.$$

If, however, the system expands from i to f and then contracts from f to i but along a different reversible transformation (different path), then the total work done would be (see figure 4.2)

$$W = \left[\int_i^f pdV \right]_1 + \left[\int_f^i pdV \right]_2$$
$$= \text{area under curve } 1 - \text{area under curve } 2$$
$$= A_{i1f2i} \neq 0$$

where A_{i1f2i} is the area enclosed by the two paths. It follows that $\oint dW = \oint pdV \neq 0$ and thus dW is not an exact differential, which means that work is not a state function. As such it depends on the particular way the system goes from i to f. Because of this from now on we will denote the incremental change in work as δW not as dW.

If we write $dV = dA ds$ (where dA is an area element and ds a distance element) we have that $\delta W = p \, dA ds = F ds = F v dt$ or that

$$\frac{\delta W}{dt} = Fv$$

or

$$\frac{\delta W}{dt} = m \frac{dv}{dt} v$$

or

$$\frac{\delta W}{dt} = \frac{d}{dt}\left(\frac{1}{2}mv^2\right)$$

or

$$\frac{\delta W}{dt} = \frac{dK}{dt} \tag{4.1}$$

where v denotes the velocity of the parcel and K is the kinetic energy of the parcel. It follows that the work done and the kinetic energy are related. The last equation indicates that one way by which a thermodynamic system can exchange energy with its environment is by performing work. The other is through transfer of heat. More on that will follow soon. According to (4.1) the units for work are those of energy. Thus, the unit for work in the MKS system is the joule which is defined as J = N m where the newton N = kg m s^{-2}. In the cgs system, the unit is the erg which is defined as erg = dyn cm where dyn = g cm s^{-2}. It follows that 1 joule = 10^7 erg.

4.2 Definition of energy

The first law of thermodynamics expresses the principle of conservation of energy for thermodynamical systems. The idea here is that energy cannot be created or destroyed. It can only change from one form to another. In this regard if during a transformation the energy of the system increases by some amount, then this amount is equal to the amount of energy the system receives from its surroundings in some other form.

Let us consider a closed system contained in an *adiabatic* enclosure. In this case the energy of the system, U, is equal to the sum of the potential and kinetic energy of all its molecules. The sum of the energies of all molecules depends on the state of the system at a given moment (i.e. on the values p, V, T) but obviously is independent of past states. It follows that the internal energy of the system depends on the state in which it exists but not on the way it arrived at that state. Thus, in a transformation $i \rightarrow f$, $\Delta U = U_f - U_i$.

Figure 4.3
The work done by an external force on a system that is taken adiabatically from a reference state O or a reference state O' to a state A. In this case the work done depends on the initial and final states, not on the particular path from an initial state to a final state.

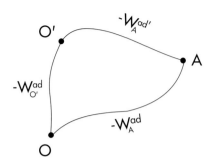

Then, for a cyclic transformation, $\oint \Delta U = 0$ which means that for an infinitesimal process dU is an exact differential. If no external forces are acting upon the system (i.e. the system is at equilibrium with its environment), then its energy remains the same ($\Delta U = 0$). If some external force acts upon the system taking it from state i to state f then the conservation of energy principle dictates that

$$U_f - U_i = -W^{\mathrm{ad}} \qquad (4.2)$$

where $-W^{\mathrm{ad}}$ is the work done adiabatically by the external force on the system ($+W^{\mathrm{ad}}$ is the work done by the system). Note that equation (4.2) implies that the work done in this case depends only on states i and f, *not* on the particular transformation $i \to f$. This is contrary to what we proved previously – that the work done is *not* an exact differential and that it depends on the particular transformation $i \to f$. The discrepancy is due to the fact that here we are dealing with adiabatic transformations. Only in this case does the work done not depend on the particular transformation.

Assume now a standard state O of this system where by definition $U_O = 0$ and a state A where $U_A \neq 0$. Assume further that through some external influences we take the system from state O to state A. Then according to equation (4.2) the energy of state A is

$$U_A = -W_A^{\mathrm{ad}}. \qquad (4.3)$$

Apparently, equation (4.3) (and hence any definition of energy) depends on the standard state O. If instead of O we had chosen another standard state O' we would have obtained U_A'. In this case (see figure 4.3) we would have (since W^{ad} depends on initial and final states only)

$$-W_A^{\mathrm{ad}} = -W_A^{\mathrm{ad}'} - W_{O'}^{\mathrm{ad}}$$

or

$$U_A = U_A' + U_{O'}$$

or

$$U_A - U_A' = U_{O'}$$

where $U_{O'}$ is the energy of the standard state O' and thus is a constant. It follows that had we chosen a different standard state, U'_A would differ by a constant (additive) amount. This constant is an essential feature of the concept of energy which has no effect on the final results when we are working with differences (changes) of energy and not with actual amounts.

4.3 Equivalence between heat and work done

Consider a sample of water at a state i where the temperature is T_i and at a state f where the temperature is $T_f > T_i$. In how many different ways can we take the system from i to f? Obviously we can simply warm up the system. In this case, and as long as T_f is low enough for evaporation to be negligible, the volume of the system remains unchanged and therefore no work is done by the external forces. This is the first way. The second way is to raise the temperature via friction. Just imagine a set of paddles rotating inside the system. In this case, we will observe that as long as the paddles continue to rotate the temperature of the water increases. Since the water resists the motion of the paddles we must perform mechanical work in order to keep the paddles moving until T_f is reached.

Thus, the work performed by the system depends on whether we go from i to f by means of the first or the second way. If we assume that the principle of the conservation of energy holds, then we must admit that the energy transmitted to the water in the form of mechanical work from the rotating paddles is transmitted to the water in the first way in a non-mechanical form called *heat*. It follows that heat and (mechanical) work are equivalent and two different aspects of the same thing: energy. In the first way, where no work is performed, the change of energy can be expressed as

$$\Delta U = Q_{W=0}$$

where $Q_{W=0}$ is the amount of heat received. In the second way (the experiment can be thought of as taking place in some adiabatic enclosure) the change in energy can be expressed as

$$\Delta U = -W^{\text{ad}}.$$

When both work and heat are allowed to be exchanged, we replace the above equations by the general expression

$$\Delta U + W = Q. \tag{4.4}$$

The quantity $Q_{W=0}$ required to raise one gram of water from 14.5 °C to 15.5 °C is by definition equal to 1 cal. In this case W^{ad} turns out to be equal to 4.185 J. This value is known as the mechanical equivalent of heat. Note that since δW is not an exact differential, δQ is not either. Thus, for an infinitesimal process the above equation takes the form

$$dU + \delta W = \delta Q. \tag{4.5}$$

Note that for a cyclic transformation $\oint dU = 0$ and thus $W = Q$. Equation (4.5) is the mathematical expression of the first law of thermodynamics. Since $\delta W = pdV$ we can write the first law as:

$$dU + pdV = \delta Q \tag{4.6}$$

or for unit mass

$$du + pda = \delta q. \tag{4.7}$$

4.4 Thermal capacities

Since dU is an exact differential we can write it as:

$$dU = \left(\frac{\partial U}{\partial T}\right)_V dT + \left(\frac{\partial U}{\partial V}\right)_T dV \tag{4.8}$$

or as:

$$dU = \left(\frac{\partial U}{\partial T}\right)_p dT + \left(\frac{\partial U}{\partial p}\right)_T dp \tag{4.9}$$

or as:

$$dU = \left(\frac{\partial U}{\partial p}\right)_V dp + \left(\frac{\partial U}{\partial V}\right)_p dV. \tag{4.10}$$

By combining equations (4.6) and (4.8) we get

$$\left(\frac{\partial U}{\partial T}\right)_V dT + \left[p + \left(\frac{\partial U}{\partial V}\right)_T\right] dV = \delta Q \tag{4.11}$$

By combining equations (4.6) and (4.9) and using the fact that

$$dV = \left(\frac{\partial V}{\partial p}\right)_T dp + \left(\frac{\partial V}{\partial T}\right)_p dT$$

we arrive at

$$\left[\left(\frac{\partial U}{\partial T}\right)_p + p\left(\frac{\partial V}{\partial T}\right)_p\right] dT + \left[\left(\frac{\partial U}{\partial p}\right)_T + p\left(\frac{\partial V}{\partial p}\right)_T\right] dp = \delta Q. \tag{4.12}$$

By combining (4.6) and (4.10) we get

$$\left(\frac{\partial U}{\partial p}\right)_V dp + \left[\left(\frac{\partial U}{\partial V}\right)_p + p\right] dV = \delta Q. \tag{4.13}$$

We define as thermal capacity, C, the ratio $\delta Q/dT$ and we denote it as C_V if the thermal capacity is measured at a constant V and as C_p if it is measured at constant p. According to this definition if in a process at constant volume there is no change of the physical and chemical state of a homogeneous system, then the heat absorbed is proportional to the variation in temperature

$$\delta Q = C_V dT$$

or

$$\delta Q = c_V m dT$$

where m is the mass of the system and c_V is the specific heat capacity at constant volume (i.e. $c_V = C_V/m$). Another specific heat is the molar specific heat $c_{Vm} = C_V/n$ where n is the number of moles. Similarly, at constant pressure

$$\delta Q = C_p dT$$

or

$$\delta Q = c_p m dT.$$

From equation (4.11) if we set $dV = 0$ we obtain

$$C_V = \frac{\delta Q}{dT} = \left(\frac{\partial U}{\partial T}\right)_V$$

and   (4.14)

$$c_V = \left(\frac{\partial u}{\partial T}\right)_a.$$

From equation (4.12) if we set $dp = 0$ we obtain

$$C_p = \left(\frac{\partial U}{\partial T}\right)_p + p\left(\frac{\partial V}{\partial T}\right)_p.$$

By defining here a new state function, the enthalpy, $H = U + pV$ ($h = u + pa$), it follows from the above equation that

$$C_p = \left(\frac{\partial H}{\partial T}\right)_p$$

and   (4.15)

$$c_p = \left(\frac{\partial h}{\partial T}\right)_p.$$

Again here we can define the molar specific heat at constant pressure as $c_{pm} = C_p/n$. Also, since $pV = nR^*T$, the enthalpy is a function of two state variables (U, T) and as such it can be expressed as an exact differential. Note that for isobaric processes $dH = dU + pdV = \delta Q$, i.e. the change in enthalpy equals the amount of heat absorbed or given away.

Figure 4.4
The experimental design
of Joule's proof that
$U = U(T)$.

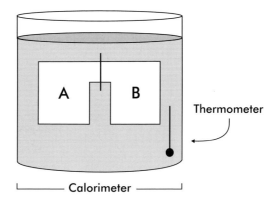

Thermometer

Calorimeter

4.5 More on the relation between U and T (Joule's law)

As we have stated earlier, any thermodynamical quantity can be
expressed as a function of two state variables. As such we can
assume that $U = f(V, T)$. In Chapter 2 we discussed the fact that
the kinetic theory of heat accepts that U is proportional to the
absolute temperature. This indicates that U is a function of only
one state variable, $U = f(T)$. An experimental proof of this was
provided by Joule, who performed the following experiment. He
constructed a calorimeter and placed in it a container having two
chambers, A and B, separated from each other by a stopcock (see
figure 4.4). He filled A with an ideal gas and evacuated B. After the
whole system settled to a thermal equilibrium (as indicated by the
thermometer in the calorimeter), Joule opened the stopcock thus
allowing the ideal gas to flow from A to B until pressure everywhere
equalized. He observed that this did not cause any change in the
temperature reading. From the first law we thus have $Q = 0$ or
from equation (4.4)

$$\Delta U = -W.$$

Now note that since the volume of the system (the container) does
not change, the system performs no work. As such $W = 0$ and
$\Delta U = 0$. But, you may ask, how about the gas inside chamber
A?. Its volume increases as it expands to fill container B. What
is the effect of that? Since there was no temperature change and
no variation in energy during the process we must conclude that
a variation in volume at *a constant temperature* produces no variation in energy. In other words $U \neq f(V)$ which means that U is
only a function of T:

$$U = U(T). \tag{4.16}$$

Later in problem 6.2 you will be asked to prove this mathematically.

More on thermal capacities

Because of equation (4.16), $(\partial U/\partial T)_V = dU/dT$. As such we can write equation (4.14) as

$$C_V = \frac{dU}{dT}$$

(4.17)

$$c_V = \frac{du}{dT}.$$

Similarly, $H = U + pV = U + nR^*T = H(T)$ and

$$C_p = \frac{dH}{dT}$$

(4.18)

$$c_p = \frac{dh}{dT}.$$

In the atmospheric range of temperatures c_V and c_p are nearly constant. From equation (4.17) we then have that

$$U = \int C_V \, dT + \text{ constant}$$

or

$$U = C_V T + \text{ constant}.$$

Since only differences in internal energy (and thus in enthalpy) are physically relevant we can set the integration constant to zero and obtain

$$U \approx C_V T.$$

Similarly $H \approx C_p T$, $u \approx c_V T$, and $h \approx c_p T$. Equations (4.17) and (4.18) can be combined to obtain

$$C_p - C_V = nR^*$$

(4.19)

which implies that

$$c_p - c_V = R, \quad c_{pm} - c_{Vm} = R^*.$$

(4.20)

Recall now from Chapter 2 (equation 2.8) that for a *monatomic* gas consisting of N point-masses the total internal energy is

$$U = \frac{3}{2} NkT.$$

It follows that

$$C_V = \frac{dU}{dT} = \frac{3}{2} Nk = \frac{3}{2} mR = \frac{3}{2} nMR = \frac{3}{2} nR^*,$$

and

$$c_{Vm} = \frac{3}{2} R^*, \quad c_V = \frac{3}{2} \frac{n}{m} R^* = \frac{3}{2} R.$$

Similarly,

$$C_p = \frac{dH}{dT} = \frac{dU}{dT} + nR^* = \frac{5}{2}nR^*,$$

and

$$c_{pm} = \frac{5}{2}R^*, \quad c_p = \frac{5}{2}R.$$

For a diatomic gas these values change to

$$C_V = \frac{5}{2}nR^*, \quad c_{Vm} = \frac{5}{2}R^*, \quad c_V = \frac{5}{2}R$$

and

$$C_p = \frac{7}{2}nR^*, \quad c_{pm} = \frac{7}{2}R^*, \quad c_p = \frac{7}{2}R.$$

Note that often in the literature the values of C_V and c_{Vm} or C_p and c_{pm} are assumed the same. This is correct only in the case of *one* mole. From the above it follows that

$$\frac{C_p}{C_V} = \frac{c_{pm}}{c_{Vm}} = \frac{c_p}{c_V}.$$

We will denote these ratios as γ.

Dry air is considered a diatomic gas. According to the above relationships we estimate that for dry air

$$\gamma_d = 1.4$$
$$c_{Vd} = 718 \text{ J kg}^{-1} \text{ K}^{-1} = 171 \text{ cal kg}^{-1} \text{ K}^{-1}$$
$$c_{pd} = 1005 \text{ J kg}^{-1} \text{ K}^{-1} = 240 \text{ cal kg}^{-1} \text{ K}^{-1}$$
$$R_d = 287 \text{ J kg}^{-1} \text{ K}^{-1}.$$

According to the above definitions we can write the first law as:

$$C_V dT + p dV = \delta Q \tag{4.21}$$

or

$$c_V dT + p da = \delta q$$

Considering that $C_V = C_p - nR^*$ and that $pV = nR^*T$, equation (4.21) becomes

$$(C_p - nR^*)dT + (nR^* dT - V dp) = \delta Q$$

or

$$C_p dT - V dp = \delta Q. \tag{4.22}$$

For convenience Table 4.1 summarizes all the different expressions of the first law.

Table 4.1. *Expressions of the first law*

For a gas of mass m	For unit mass
$dU + \delta W = \delta Q$	$du + \delta w = \delta q$
$dU + pdV = \delta Q$	$du + pda = \delta q$
$C_V dT + pdV = \delta Q$	$c_V dT + pda = \delta q$
$C_p dT - V dp = \delta Q$	$c_p dT - adp = \delta q$

4.6 Consequences of the first law

Let us look at the form of the first law in the following special cases.

- Isothermal transformations: $i \xrightarrow{T=\text{constant}} f$.

During such transformations $dT = 0$ and as such from equation (4.21) it follows that $\delta Q = \delta W$. Then the amount of heat exchanged is

$$Q = \int_i^f \delta W = \int_i^f pdV$$

or

$$Q = nR^*T \int_i^f \frac{dV}{V}$$

or

$$Q = nR^*T \, \ln \frac{V_f}{V_i}.$$

- Isochoric transformations: $i \xrightarrow{V=\text{constant}} f$.

In this case $dV = 0$ and from the 1st law it follows that

$$\delta Q = dU = C_V dT$$

or

$$\Delta U = Q = C_V \int_i^f dT = C_V (T_f - T_i).$$

- Isobaric transformations: $i \xrightarrow{p=\text{constant}} f$.

In this case it follows that

$$\delta Q = C_p dT \quad \text{or} \quad Q = C_p(T_f - T_i),$$

$$\delta W = pdV \quad \text{or} \quad W = p(V_f - V_i).$$

and

$$\Delta U = C_p(T_f - T_i) - p(V_f - V_i)$$

- Cyclic transformations $i \longrightarrow f \longrightarrow i$.

As we know, in cyclic transformations $\oint dV = 0$. Then from the 1st law it follows that

$$\oint \delta W = \oint \delta Q.$$

- Adiabatic transformations – Poisson's relations.

During an adiabatic process there is no exchange of heat between a system and its surroundings. Thus, $\delta Q = 0$. Then the 1st law can be expressed as

$$dU = -\delta W$$

or

$$C_V dT = -p dV$$

or

$$C_p dT = V dp$$

or

$$\frac{dT}{T} = \frac{V}{C_p} \frac{dp}{T}$$

or using $pV = mRT$

$$\frac{dT}{T} = \frac{mR}{C_p} \frac{dp}{p}$$

or

$$\frac{dT}{T} = \frac{nR^*}{C_p} \frac{dp}{p}$$

or using $C_p - C_V = nR^*$

$$\frac{dT}{T} = \left(1 - \frac{C_V}{C_p}\right) \frac{dp}{p}$$

or

$$\frac{dT}{T} = \left(1 - \frac{1}{\gamma}\right) \frac{dp}{p}. \qquad (4.23)$$

Integrating equation (4.23) yields

$$\ln T = \left(1 - \frac{1}{\gamma}\right) \ln p + \ln(\text{constant})$$

or

$$\ln T = \ln p^{\frac{\gamma - 1}{\gamma}} + \ln(\text{constant})$$

Figure 4.5
The various relationships between adiabats, isotherms, isobars, and isochores in partial and in the complete state spaces.

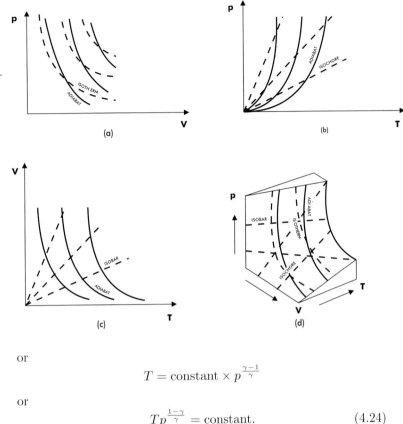

or

$$T = \text{constant} \times p^{\frac{\gamma-1}{\gamma}}$$

or

$$Tp^{\frac{1-\gamma}{\gamma}} = \text{constant}. \tag{4.24}$$

Through the ideal gas law we can then derive the following equivalent expressions of equation (4.24):

$$TV^{\gamma-1} = \text{constant} \tag{4.25}$$

$$pV^{\gamma} = \text{constant}. \tag{4.26}$$

Equations (4.24)–(4.26) express Poisson's relations for adiabatic processes. Since for dry air $\gamma = 1.4 > 0$ equation (4.25) proves something that we all more or less know: that as an air parcel rises and expands (i.e. V increases), its temperature must decrease. For adiabatic expansion, since δQ is zero this decrease is due to expansion work that the parcel performs on the environment.

From the ideal gas law it follows that isotherms are given by the equation

$$pV = \text{constant}.$$

Thus on a (p, V) diagram isotherms are equilateral hyperbolas (figure 4.5(a); an equilateral hyperbola is one whose asymptotes are perpendicular to each other). From equation (4.26) we see that the adiabats on a (p, V) diagram will also be equilateral hyperbolas but since $\gamma > 1$ the adiabats will be steeper than the isotherms. Obviously in a (p, V) diagram isobaric and isochoric transformations are represented by straight lines on the p and V axis, respectively. On a (p, T) diagram the isochores are given by the equation

$$V = \frac{nR^*T}{p} = \text{constant}'$$

or

$$Tp^{-1} = \text{constant}$$

which is the equation of a straight line. From equation (4.24) it also follows that in a (p, T) diagram the adiabats are again hyperbolas (figure 4.5(b)). On a (T, V) diagram (figure 4.5(c)) the isobars are straight lines (as before $TV^{-1} = \text{constant}$) and the adiabats are hyperbolas (see equation 4.25). The complete (p, V, T) space and transformations are shown in figure 4.5(d).

- **Polytropic transformations**

  During an isothermal transformation the system can exchange heat with its environment but its temperature remains constant. This, however, is an ideal situation. In reality the equilization of temperatures does not happen instantaneously. Similarly, an adiabatic transformation is an ideal situation since it requires perfect heat insulation. The same comments can be made for isobaric and isochoric transformations. Thus, the relations $pV = \text{constant}$, $pV^\gamma = \text{constant}$, $Tp^{-1} = \text{constant}$, and $TV^{-1} = \text{constant}$ are special cases of the general relationship $pV^\eta = \text{constant}$ where η can take on any value. Such transformations are called polytropic transformations (in Greek polytropic means multi-behavioral). If we assume that $\eta = \gamma/\epsilon$ where ϵ is a constant, we can work our way backwards from equation $pV^\eta = \text{constant}$ to arrive at

$$\frac{dT}{T} = \left(1 - \frac{\epsilon}{\gamma}\right)\frac{dp}{p}$$

  and subsequently at

$$C_p dT = V dp - \frac{C_p(\epsilon - 1)}{\gamma}\frac{V}{nR^*}dp.$$

  Then, according to the first law, the second term on the right-hand side is the heat exchanged, which is zero if $\epsilon = 1$ (i.e. $\eta = \gamma$, adiabatic process), greater than zero (i.e. the system absorbs heat from the environment) if $\eta < \gamma$ ($\epsilon > 1$) and $dp < 0$, or $\eta > \gamma$ ($\epsilon < 1$) and $dp > 0$, and less than zero (i.e. the system gives away heat to the environment) if $\eta > \gamma$ and $dp < 0$, or $\eta < \gamma$ and $dp > 0$. From the above equation we also find that for $\epsilon = 0$ (i.e. $\eta = \infty$) $nR^* dT = V dp$, which owing to the ideal gas law implies $dV = 0$. For $\epsilon = \gamma$ (i.e. $\eta = 1$), it reduces to

$C_p dT = 0$, while for $\epsilon = \infty$ (i.e. $\eta = 0$) it dictates that $dp = 0$. As such a polytropic process $pV^\eta = $ constant reduces to an isobaric, to an isothermal, to an adiabatic, and to isochoric process for $\eta = 0, 1, \gamma$, and ∞, respectively.

Having said this we must add that in the atmosphere over a rather wide range of motions the timescale for an air parcel (our system) needed to adjust to environmental changes of pressure, and to perform work, is short compared with the corresponding timescale of heat transfer. For example, above the boundary layer and outside the clouds the timescale of heat transfer is about two weeks whereas the timescale of displacements that affect a parcel is of the order of hours to a day. It thus appears that adiabatic approximations are good approximations for many atmospheric phenomena.

- **Dry adiabatic lapse rate**
From equation (4.24) we have

$$T = \text{constant} \cdot p^{\frac{\gamma - 1}{\gamma}}.$$

By taking the logarithmic derivative of the above equation we have

$$d \ln T = \frac{\gamma - 1}{\gamma} d \ln p$$

or

$$\frac{dT}{T} = \frac{\gamma - 1}{\gamma} \frac{dp}{p}$$

or

$$\frac{1}{T} \frac{dT}{dz} = \frac{\gamma - 1}{\gamma} \frac{1}{p} \frac{dp}{dz}. \tag{4.27}$$

For large-scale motions the hydrostatic approximation states that pressure gradient force balances the force due to gravity. Therefore,

$$\frac{dp}{dz} = -\rho g. \tag{4.28}$$

For a parcel of air rising in the atmosphere, equation (4.27) is valid as long as its ascent is an adiabatic process. For the same parcel equation (4.28) is valid only if dp/dz experienced by the parcel is equal to that of the large scale (surroundings). We assume this is more or less true but in equation (4.28) ρ represents the density of the surroundings which from the ideal gas law is

$$\rho = \frac{p}{RT_S} \tag{4.29}$$

where T_S denotes the temperature of the surroundings. In this case by combining equations (4.27)–(4.29) we have that

$$\frac{dT}{dz} = -\frac{\gamma - 1}{\gamma}\frac{g}{R}\frac{T}{T_S}.$$

If the parcel remains dry (i.e. no condensation takes place) then from the above equation we may define the dry adiabatic lapse rate, Γ_d, as the atmospheric temperature profile such that the temperature of the parcel is always at the temperature of its surroundings. This guarantees that the process undergone by the parcel (ascent or descent) is adiabatic. For such a profile $T = T_S$ and as such for dry air

$$\Gamma_d = -\frac{dT}{dz} = \frac{\gamma - 1}{\gamma}\frac{g}{R_d} = \frac{g}{c_{pd}} = 9.8\ ^\circ\text{C km}^{-1}. \tag{4.30}$$

This value is larger than the observed average decrease of temperature with altitude. The difference is mainly due to the fact that in the derivation of Γ_d we neglected the effect of condensation of water vapor. When this happens we have a new lapse rate which we will discuss in a later chapter.

Recalling that $h = u + pa$ and that $h = c_{pd}T$ we can write equation (4.30) as

$$\frac{dh}{dz} = -g$$

or

$$\frac{d}{dz}(h + gz) = 0 \tag{4.31}$$

where $h + gz$ is defined as the dry static energy. The above equations indicate (1) that enthalpy of a parcel decreases as it rises adiabatically because it is doing gravitational work, and (2) that the static energy (the sum of enthalpy and the gravitational potential energy) is conserved in an adiabatic motion (ascent or descent). Enthalpy is thus the specific internal energy plus the term pa that accounts for the work done by the parcel on the surroundings.

- **Potential temperature**
  Let us assume that there is a variable θ defined by the relation $\theta = ATp^{-\beta}$ where A and β are constants. By taking the logarithm on both sides we get $\ln\theta = \ln A + \ln T - \beta\ln p$. Then by differentiating we arrive at

$$d\ln\theta = d\ln T - \beta d\ln p.$$

Consider now the first law in the form $c_p dT - adp = \delta q$ and divide it by T. Then with the help of the ideal gas we obtain

$$d\ln T - \frac{R}{c_p}d\ln p = \frac{\delta q}{c_p T}.$$

For $\beta = R/c_p$ the above two equations combine to give

$$d \ln \theta = \frac{\delta q}{c_p T}. \tag{4.32}$$

For adiabatic processes $\delta q = 0$ and thus $d \ln \theta = 0$. As such we should expect that in adiabatic processes there should exist a measure θ which is conserved. This measure can be defined as follows.

From equation (4.24) we have

$$\frac{T}{p^k} = \frac{T_0}{p_0^k}$$

where $k = (\gamma - 1)/\gamma = 1 - c_V/c_p = R/c_p = 0.286$ (for dry air). The state (T_0, p_0) can be taken as a reference state. As such we can choose $p_0 = 1000$ mbar and denote the corresponding temperature as T_0. The above equation is expressed as

$$T_0 = T \left(\frac{p_0}{p} \right)^k$$

or $\tag{4.33}$

$$T_0 = p_0^k T p^{-k}.$$

If we compare the above expression to $\theta = AT p^{-\beta}$ we see that $T_0 = \theta$ for $A = p_0^k$ and $\beta = k$. We call θ the potential temperature and we regard it as the temperature a parcel will have if it is compressed or expanded *adiabatically* from any state (T, p) to the 1000 mbar level. It follows that the potential temperature remains invariant during an adiabatic process. Thus, under adiabatic conditions θ can be used as a tracer of air motion. On timescales for which a parcel can be considered as adiabatic, constant values of θ track the motion of this parcel, which takes place on a surface of constant θ. From equation (4.33) it is clear that the distribution of θ in the atmosphere depends on the distribution of T and p. In the atmosphere $dp/dz \gg dT/dz$ (just think that 5 km above the surface pressure decreases on the average from about 1000 mbar to 500 mbar while the temperature decreases from 288 K to 238 K). Therefore, surfaces of constant θ tend to be like isobaric surfaces. Before, we saw that for an ascent (descent) the parcel's temperature must decrease (increase) owing to the work done by the parcel (by the surroundings). With the definition of potential temperature we can extend this statement to read as follows: during an adiabatic ascent (descent) the parcel's temperature must decrease (increase) but in such a proportion as to preserve the parcel's potential temperature. Note that equation (4.32) indicates that for non-adiabatic processes the change in the potential energy is a direct measure of the heat exchanged between the parcel and its environment. As a result a non-adiabatic parcel will drift across potential temperature surfaces in proportion to the net amount of heat exchanged with its environment.

Examples

(4.1)   One mole of a gas expands from a volume of 10 liters and a temperature of 300 K to (a) a volume of 14 liters and a temperature of 300 K and (b) a volume of 14 liters and a temperature of 290 K. What is the work done by the gas on the environment in each case?

(a)   According to the definition of work

$$W = \int_{V_1}^{V_2} pdV = \int_{V_1}^{V_2} \frac{nR^*T}{V} dV.$$

Since the temperature remains the same the above equation gives

$$W = nR^*T \ln \frac{V_2}{V_1} = 839 \text{ J.}$$

(b)   In this case T does not remain constant. As such it cannot be taken outside the integral and the above equation cannot be applied.

When T does not remain constant, in order to calculate the work done we must have the explicit function describing the path from i to f in the (p, V) diagram, not just i and f. An approximation to this would be to define a temperature $\overline{T}$ or a pressure $\overline{p}$ that satisfied the relationship

$$\int_{V_1}^{V_2} \frac{TdV}{V} \approx \overline{T} \int_{V_1}^{V_2} \frac{dV}{V}$$

or the relationship

$$\int_{V_1}^{V_2} pdV = \overline{p} \int_{V_1}^{V_2} dV$$

Since in our case no clues are given about the function describing the path from i to f we are free to make (reasonable) assumptions. We can thus assume that i and f are connected by a straight line. From the data we can estimate that $p_i = 249\,420$ Pa and and $p_f = 172\,200$ Pa. From figure 4.6 it is easy to see that the work done is the area of the trapezoid ifV_fV_i. Because of the equality of the shaded triangles this area is equal to the area of the rectangle defined by the lines $p = 0$, $p = \overline{p}$, $V = V_i$, $V = V_f$ where $\overline{p} = p_1 + p_2/2$. Thus

$$W = \overline{p}(V_f - V_i) = 843 \text{ J.}$$

Figure 4.6
Diagram for example 4.1.

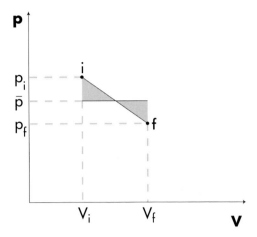

Figure 4.7
Diagram for example 4.2.

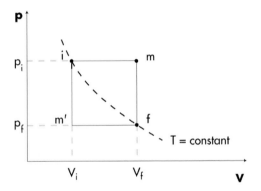

(4.2)   Consider an initial, i, and a final, f, state of a gas on an isotherm with $V_f > V_i$. The gas goes from i to f via an isobaric expansion and subsequent isochoric cooling; sketch on a (p, V) diagram the complete transformation and find the work done. If instead the gas goes from i to f via an isochoric cooling followed by an isobaric expansion, will the work done be different from before?

In the first case the gas goes from i to m (isobaric from V_i to V_f) and then from m to f (isochoric from p_i to p_f) (figure 4.7). The total work done from $i \rightarrow f$ will be the sum of the work done from $i \rightarrow m$ and from $m \rightarrow f$

$$W_1 = \int_{V_i}^{V_f} p dV + \int_{V_f}^{V_f} p dV$$

$$= p_i \int_{V_i}^{V_f} dV = p_i(V_f - V_i).$$

In the second case the gas goes from i to m' (isochoric from p_i to p_f) and then from m' to f (isobaric from V_i to V_f).

As before we have that the total work from $i \to f$ is

$$W_2 = \int_{V_i}^{V_i} p\,dV + \int_{V_i}^{V_f} p\,dV$$

$$= p_f \int_{V_i}^{V_f} dV = p_f (V_f - V_i).$$

Since $p_i \neq p_f$ it follows that $W_1 \neq W_2$.

(4.3) Calculate Q, W, and ΔU for the following processes:

(a) isothermal reversible compression from state $i = (p_i, V_i)$ to state $f = (p_f, V_f)$;

(b) adiabatic reversible compression from state $i = (p_i, V_i)$ to state $m = (p_f, V_m)$ and a subsequent isobaric reversible compression to state $f = (p_f, V_f)$;

(c) reversible isochoric increase of the temperature from state $i = (p_i, V_i)$ to state $m' = (p_f, V_i)$ and a subsequent reversible isobaric decrease in temperature to state $f = (p_f, V_f)$.

Express all answers in terms of p_i, p_f and T, where T is the temperature of the isotherm in (a).

(a) During this transformation $T_i = T_f = T, \Delta T = 0$ and $\Delta U = 0$. Then from the first law it follows that

$$Q = W = \int_i^f p\,dV = nR^*T \int_i^f \frac{dV}{V}$$

$$= nR^*T \ln \frac{V_f}{V_i} = nR^*T \ln \frac{p_i}{p_f}.$$

(b) This transformation consists of two branches $(i \to m$ and $m \to f)$. From (a), the initial and final temperature is the same, i.e. $\Delta U = 0$. Thus, again we have $Q = W$ with

$$\Delta U = \Delta U_{i \to m} + \Delta U_{m \to f}$$

$$Q = Q_{i \to m} + Q_{m \to f}$$

and

$$W = W_{i \to m} + W_{m \to f}.$$

Since the branch $i \to m$ is adiabatic it follows that $Q_{i \to m} = 0$ and

$$W_{i \to m} = -\Delta U_{i \to m} = -C_V (T_m - T_i). \qquad (4.34)$$

T_m and T_i are related via the equation

$$T_i p_i^{\frac{1-\gamma}{\gamma}} = T_m p_m^{\frac{1-\gamma}{\gamma}},$$

where $p_m = p_f$. It follows that

$$T_m = T_i \left(\frac{p_i}{p_f}\right)^{\frac{1-\gamma}{\gamma}} = T \left(\frac{p_i}{p_f}\right)^{\frac{1-\gamma}{\gamma}}.$$

Then from equation (4.34) we obtain

$$W_{i \to m} = -C_V T \left[\left(\frac{p_i}{p_f}\right)^{\frac{1-\gamma}{\gamma}} - 1\right]. \qquad (4.35)$$

Branch $m \to f$ is isobaric. Thus,

$$W_{m \to f} = \int_m^f p \, dV = p_f(V_f - V_m). \qquad (4.36)$$

V_m and V_i are related via Poisson's equation as follows

$$p_i V_i^\gamma = p_m V_m^\gamma.$$

Since $p_m = p_f$ the above equation becomes

$$\left(\frac{V_m}{V_i}\right)^\gamma = \frac{p_i}{p_f}$$

or

$$V_m = V_i \left(\frac{p_i}{p_f}\right)^{1/\gamma}.$$

Then equation (4.36) can be written as

$$W_{m \to f} = p_f \left[V_f - V_i \left(\frac{p_i}{p_f}\right)^{1/\gamma}\right]$$

$$= p_f V_f \left[1 - \frac{V_i}{V_f} \left(\frac{p_i}{p_f}\right)^{1/\gamma}\right]$$

$$= nR^*T \left[1 - \frac{p_f}{p_i} \left(\frac{p_i}{p_f}\right)^{1/\gamma}\right]$$

$$= nR^*T \left[1 - \left(\frac{p_i}{p_f}\right)^{\frac{1-\gamma}{\gamma}}\right] \qquad (4.37)$$

From equations (4.35) and (4.37) it follows that overall

$$Q = W = -C_V T \left[\left(\frac{p_i}{p_f}\right)^{\frac{1-\gamma}{\gamma}} - 1\right]$$

$$+ nR^*T \left[1 - \left(\frac{p_i}{p_f}\right)^{\frac{1-\gamma}{\gamma}}\right]$$

$$= C_p T \left[1 - \left(\frac{p_i}{p_f}\right)^{\frac{1-\gamma}{\gamma}}\right].$$

Figure 4.8
Diagram for example 4.3.

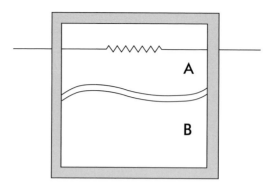

(c)   Here also the transformation consists of two branches, $i \to m'$ and $m' \to f$. Branch $i \to m'$ is isochoric, so

$$W_{i \to m'} = 0.$$

Branch $m' \to f$ is isobaric, which gives

$$W_{m' \to f} = \int_{m'}^{f} p dV = p_f(V_f - V_{m'}) = p_f(V_f - V_i)$$

or

$$W_{m' \to f} = p_f V_f \left(1 - \frac{V_i}{V_f}\right)$$

$$= nR^* T \left(1 - \frac{p_f}{p_i}\right)$$

Here again $T_i = T_f = T$, $\Delta T = 0$ and $\Delta U = 0$. Thus, overall

$$Q = W = nR^* T \left(1 - \frac{p_f}{p_i}\right).$$

(4.4)   This following problem is due to Iribarne and Godson (1973). Students should try to solve it before looking at the solution here. Figure 4.8 shows an insulated box with two compartments A and B. Both compartments contain a monatomic ideal gas and they are separated by a wall that on one hand does not allow any heat through it but on the other hand is flexible enough to ensure equalization of pressure in both compartments. The initial conditions for both compartments are the same: T_i = 273 K, V_i = 1000 cm³, and p_i = 1 atmosphere (1013 × 10² Pa). Then, by means of an electrical resistance, heat is supplied to the gas in A until its pressure becomes ten times its initial pressure. Estimate: (a) the final temperature of the gas in B, (b) the work performed on the gas in B, (c) the final temperature of the gas in A, and (d) the heat absorbed by the gas in A.

(a) The heat that is supplied to gas in A results in an increase in all its state variables. As it expands, it com-
presses the gas in B but it does not provide it with
any of its heat because of the insulating wall between
them. Since the wall is flexible the gas in B is compressed adiabatically until its pressure equals that inside compartment A. As such $p_{Bf} = 10$ atmospheres.
From Poisson's equation we can then solve for T_{Bf}

$$T_{Bi}\, p_{Bi}^{\frac{1-\gamma}{\gamma}} = T_{Bf}\, p_{Bf}^{\frac{1-\gamma}{\gamma}}$$

Since the gas is monatomic $\gamma = 1.666$. Thus,

$$T_{Bf} = \left(\frac{p_{Bi}}{p_{Bf}}\right)^{-0.4} T_{Bi} \approx 686 \text{ K}.$$

Having T_{Bf} we can use the other form of Poisson's
equation to find V_{Bf}:

$$T_{Bi}V_{Bi}^{\gamma-1} = T_{Bf}V_{Bf}^{\gamma-1}$$

or

$$V_{Bf} = \left(\frac{T_{Bi}}{T_{Bf}}\right)^{1/\gamma-1} V_{Bi} = \left(\frac{273}{686}\right)^{1.5} V_{Bi}$$
$$= 0.25\, V_{Bi} = 250 \text{ cm}^3.$$

(b) The work performed on the gas in B is given by

$$W = -\Delta U = -C_V(T_{Bf} - T_{Bi}) = -413\, C_V \text{ J}.$$

As we know, for a monatomic gas $C_V = \frac{3}{2}nR^*$. The
number of moles n can be found from the ideal gas law
and the initial conditions. It follows that

$$n = \frac{p_i V_i}{R^* T_i} = 0.0446 \text{ mol}.$$

Accordingly, $C_V = 0.556$ JK^{-1} and

$$W \approx -230 \text{ J}$$

or

$$W \approx -55 \text{ cal}.$$

(c) Since the final volume in B is 250 cm^3 it follows that
the final volume in A is 1750 cm^3. By applying the
ideal gas law for the initial and final conditions in A
we find that

$$T_{Af} = \frac{p_{Af}V_{Af}}{p_{Ai}V_{Ai}}T_{Ai} = 4777 \text{ K}.$$

(d)   The heat absorbed by the gas in A is given from the first law

$$Q = \Delta U + W.$$

Here $W = 55$ cal (i.e. the opposite of the work done *on* gas B), and $\Delta U = C_V(T_{Af} - T_{Ai}) = 2504$ J $= 598$ cal. Thus

$$Q = 598 + 55 = 653 \text{ cal.}$$

Problems

(4.1)   Commercial aircraft fly near 200 mbar where typically the outside temperature is $-60\,°$C. (a) Calculate the temperature of air if compressed adiabatically to the cabin pressure of 1000 mbar. (b) How much specific heat must be added or removed to maintain the cabin at $25\,°$C? Consider the air as dry air. (337.5 K, 9.5 cal g^{-1} has to be removed)

(4.2)   A sample of 100 grams of dry air has an initial temperature of 270 K and pressure 900 mbar. During an isobaric process heat is added and the volume expands by 20% of its initial volume. Estimate: (a) the final temperature of the air, (b) the amount of heat added, and (c) the work done against the environment. (324 K, 5427 J, 1550 J)

(4.3)   An air parcel moves from p_i to p_f. If its initial temperature is T_i find: (a) the specific work done by or on the parcel if its change happens adiabatically, (b) the specific work if its change happens isothermally, and (c) the change in the parcel's potential temperature in (a) and (b).

(4.4)   One mole of dry air has an initial state $T = 273$ K and $p = 1$ atmosphere. It undergoes a process in which its volume becomes four times its initial volume at 400 mbar. If air is considered an ideal gas and if the process obeys the law $pV^\eta = $ constant, estimate (a) the value of η, (b) the final temperature, (c) the change in internal energy, the work done, and the heat exchanged between the air and its surroundings. Look at your results carefully and elaborate on what kind of process would generate results like these. Is this process a realistic process? (0.67, 431 K, 3286 J, 3989 J, 7275 J)

(4.5)   Assume that you are on the top of a mountain at an altitude of 15 000 feet and there are no clouds above or below. If the temperature is $-12\,°$C what will the temperature be at 3500 feet altitude? ($22.4\,°$C)

(4.6)   A parcel of dry air has a volume 10 liters, a temperature

Figure 4.9
Diagram for example 4.9.

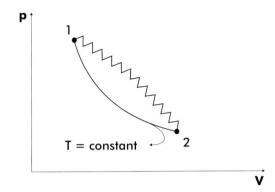

27 °C, and a pressure 1 atmosphere. The parcel (a) is compressed isothermally until its volume becomes 2 liters, and (b) expands adiabatically until its volume becomes 10 liters. Describe these changes graphically in a (p, V) diagram. On it mark the values of p, V, T for each case.

(4.7)   A sample of dry air has a temperature of 300 K, a volume of 3 liters and a pressure of 4 atmospheres. The air then undergoes the following changes: (a) it is warmed under a constant pressure to 500 K, (b) it is cooled under constant volume to 250 K, (c) it is cooled under constant pressure to 150 K, and (d) it is warmed under constant volume to 300 K. (1) Describe graphically on a (p, V) diagram each of the changes and mark on it the final values for pressure and volume after each change. (2) Calculate the total work done. (405.2 J)

(4.8)   For an ideal gas what on a (p, V) diagram describes the lines for which $U = $ constant?

(4.9)   An ideal gas undergoes two transformations from state 1 to state 2 which follow the two curves in figure 4.9. In which of the transformations is (a) the change in internal energy greater? (b) the heat absorbed greater?

(4.10) One kilogram of dry air is warmed from 20 °C to 70 °C under a constant pressure of 1 atmosphere. Calculate: (a) the heat absorbed by the gas, (b) the work done, and (c) the change in internal energy. (50 250 J, 14 350 J, 35 900 J)

(4.11) Hydrogen under $p = 5$ atmospheres and $T = 20$ °C is warmed isobarically until its volume increases from 1 to 2 liters. Calculate the absorbed heat. (1773.4 J)

(4.12) Inside a cylinder closed by a piston there is 88 grams of an unknown diatomic gas of a temperature 0 °C. The gas is then compressed adiabatically until its volume becomes equal to 1/10 of the initial volume. The change in internal energy is 17 158 J. Identify the gas. (Carbon dioxide)

(4.13) Provide an example that proves that heat transfer is an irreversible process.

(4.14) At $T = 0\,^\circ\text{C}$ and $p = 1000$ mbar one gram of dry air receives 5 calories of heat. It is then observed that the pressure drops 50 mbar. What is the change in the temperature of the air? $(16.9\,^\circ\text{C})$

(4.15) Calculate the change in specific energy of an adiabatic parcel whose speed changes from 10 m s^{-1} to 25 m s^{-1}. $(-262.5 \text{ J kg}^{-1})$

CHAPTER FIVE

The second law of thermodynamics

The first law of thermodynamics arose from the conservation of energy principle. The first law, even though it implies that we cannot create or destroy energy, places no limits on how energy can be transformed from one form to another. Thus, on the basis of the first law, heat can be transformed into work, work into heat, work can be done at the expense of internal energy, and so on. However, if no other laws existed the first law would allow phenomena to happen that never happen in reality. For example, consider a heavy body falling on the ground. We will observe that during the impact the body will warm. The opposite phenomenon according to which a body at rest on the ground begins to rise by itself while it is cooling is impossible. Similarly, no engine has yet been built which, for instance, would receive heat from the sea, transform it to work, and then set a ship in motion. Both the above examples are not in disagreement with the first law since the work would be done at the expense of the internal energy of the soil or the sea. The impossibility of these phenomena is due to the second law of thermodynamics, often hailed as the supreme law of nature. We will start our discussion of this law with the following example.

5.1 The Carnot cycle

The Carnot cycle is a thermal engine. A thermal engine is one that receives from some source an amount of heat, part of which it transforms into work. In thermodynamical terms the Carnot cycle is a cyclic transformation consisting of the following four steps (figure 5.1):

Step 1: reversible isothermal expansion $(1 \longrightarrow 2)$, $T = T_1 = $ constant

Figure 5.1
Steps in a Carnot cycle.

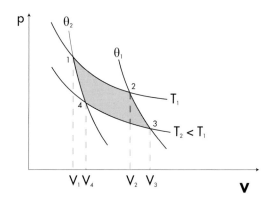

Step 2: reversible adiabatic expansion $(2 \longrightarrow 3)$, $\theta = \theta_1 =$ constant

Step 3: reversible isothermal compression $(3 \longrightarrow 4)$, $T = T_2 =$ constant

Step 4: reversible adiabatic compression $(4 \longrightarrow 1)$, $\theta = \theta_2 =$ constant

with $T_2 < T_1$, $\theta_2 < \theta_1$.

Let us now calculate the work done and heat absorbed when an ideal gas is subjected to this cycle. The change in internal energy is zero because the overall transformation is a cyclic one. We will consider each step separately.

Step 1

$$\Delta T = 0$$
$$W_{12} = \int_{V_1}^{V_2} pdV = nR^*T_1 \int_{V_1}^{V_2} \frac{dV}{V} = nR^*T_1 \ln \frac{V_2}{V_1}$$
$$\Delta U_{12} = C_V \Delta T = 0$$
$$Q_{12} = W_{12}$$

Since $V_2 > V_1$ it follows that $W_{12}, Q_{12} > 0$ indicating that during step 1 work was done by the gas and heat was absorbed by the gas from the heat source at temperature T_1.

Step 2

$$Q_{23} = 0$$
$$\Delta U_{23} = C_V(T_2 - T_1) < 0$$
$$W_{23} = -\Delta U_{23} = -C_V(T_2 - T_1) > 0$$

Step 3

$$\Delta T = 0$$
$$W_{34} = \int_{V_3}^{V_4} pdV = nR^*T_2 \ln \frac{V_4}{V_3}$$
$$\Delta U_{34} = 0$$
$$Q_{34} = W_{34}$$

Since $V_4 < V_3$ it follows that $W_{34}, Q_{34} < 0$ indicating that now work is done on the gas and heat is given away by the gas to the source at temperature T_2.

Step 4

$$Q_{41} = 0$$
$$\Delta U_{41} = C_V(T_1 - T_2) > 0$$
$$W_{41} = -\Delta U_{41} = -C_V(T_1 - T_2) < 0$$

From the above it follows that the total work done is

$$W = W_{12} + W_{23} + W_{34} + W_{41}$$
$$= nR^*T_1 \ln \frac{V_2}{V_1} + nR^*T_2 \ln \frac{V_4}{V_3} \qquad (5.1)$$

Because transformations $2 \to 3$ and $4 \to 1$ are adiabatic the following relations apply:

$$T_1 V_2^{\gamma-1} = T_2 V_3^{\gamma-1}$$

and

$$T_1 V_1^{\gamma-1} = T_2 V_4^{\gamma-1}.$$

By dividing the above two equations we have

$$\frac{V_2}{V_1} = \frac{V_3}{V_4}.$$

Accordingly, equation (5.1) can be written as

$$W = nR^*T_1 \ln \frac{V_2}{V_1} - nR^*T_2 \ln \frac{V_2}{V_1}$$

or

$$W = nR^* \left(\ln \frac{V_2}{V_1} \right) (T_1 - T_2).$$

The total amount of heat absorbed during this cycle is

$$Q = Q_{12} + Q_{34} = W_{12} + W_{34} \qquad (5.2)$$

or

$$Q = W = nR^* \left(\ln \frac{V_2}{V_1} \right) (T_1 - T_2) \qquad (5.3)$$

For simplicity we will now denote the amount of heat absorbed by the gas from the heat source at T_1 (i.e. Q_{12}) as Q_1 and the amount of heat given by the gas to the source at T_2 (i.e. Q_{34}) as Q_2. Since W is the area enclosed by the cycle it follows that $W > 0$. Then from equation (5.3) it follows that $Q > 0$. Since $W = Q_1 + Q_2$ and Q_2 is negative it follows that only part of the heat absorbed by the gas at the source T_1 (higher temperature) is transformed into work. The other part is surrendered to the source at T_2 (lower temperature). We define the efficiency of the Carnot cycle, η, as

the ratio between the work done and the heat absorbed by the gas
at the source T_1.

$$\eta = \frac{Q_1 + Q_2}{Q_1} = 1 + \frac{Q_2}{Q_1} \quad (Q_1 > 0, Q_2 < 0).$$

By considering the relations derived for Q_1 and Q_2 we can write
the above ratio as

$$\eta = 1 + \frac{nR^* T_2 \ln \frac{V_4}{V_3}}{nR^* T_1 \ln \frac{V_2}{V_1}}$$

or

$$\eta = 1 - \frac{T_2}{T_1} \tag{5.4}$$

5.2 Lessons learned from the Carnot cycle

- According to equation (5.4) the thermodynamic efficiency of the
  cycle depends only on the temperatures of the two heat sources
  and it becomes zero if $T_2 = T_1$ (i.e. if the two sources collapse into
  one source). As such we conclude that doing work via a thermal
  engine operating at only one source of heat is impossible. This is
  known as Kelvin's postulate and is an expression of the second
  law of thermodynamics. Another way of stating this postulate
  is that it is impossible to construct an engine which transforms
  heat into work without surrendering some heat to a source at a
  lower temperature. This explains why stones on the ground do
  not rise in the air or why no machine has been discovered which
  by taking heat from the sea or the air sets in motion a ship or
  a car. According to the second law, apart from the warm heat
  source (sea or air), a second colder heat source would be required
  which we do not really have. I use the word *really* because some
  differences in temperature do exist in both the oceans and the
  atmosphere. For example, the temperature of the surface air is
  around 35 K warmer than the air at 5 km, and the temperature
  at the surface of the oceans is about 20 K warmer than at a depth
  of 1 km. In both cases, however, it turns out that the Carnot
  efficiency is about 10%. This is a low efficiency given the tremendous expenses that would be required to build such engines. On
  the other hand if the atmosphere, in the horizontal, is thought
  of as a Carnot cycle operating at $T_1 =$ equatorial temperatures
  and $T_2 =$ polar temperatures we can explain why in the winter
  the circulation is stronger. Since T_1 is approximately constant
  throughout the year, while T_2 varies significantly, it follows that
  the efficiency of this "machine" is higher in the winter than in
  the summer. Accordingly, the amount of heat given by the atmosphere to the cold source is smaller in the winter, than in the

summer. Since the amount of heat absorbed by the atmosphere from the warm source is more or less constant, this means that there is more heat "remaining" in the atmosphere in the winter than in the summer that can be transformed into kinetic energy.

Apart from the impossibility of transformations whose only final result is to transform into work heat extracted from only one source of heat, there is another type of transformation that is impossible. Such transformations refer to transfer of heat from a cold body to a warm body without work being done. This phenomenon is allowed by the first law (since heat lost by the cold body will be exactly equal to the heat gained by the warm body) but has never been observed. Of course transformations that result in heat flow from a colder body to a warmer body are allowed if work is done (such as in refrigerators). This leads to another expression of the second law of thermodynamics. A transformation whose only final result is to transfer heat from a body at a given temperature to a body at a higher temperature is impossible. This is known as Clausius's postulate. It is easy to show that both postulates are equivalent and two expressions of the same law. To do that we shall prove that if Clausius was wrong then Kelvin would be wrong and vice versa. If Kelvin was wrong and we could transform into work heat extracted from only one heat source, then we could transform this work (by means of friction) into heat and with this heat raise the temperature of a given body of a higher initial temperature. This would be a violation of Clausius's postulate. Now if Clausius was wrong and we could transfer a certain amount Q_1 from a source at T_2 to a source at T_1 ($T_2 < T_1$) in such a way that no other change in the state of the system occurred, then with the help of a Carnot cycle we could absorb this amount and produce an amount of work, W. Since the source T_1 receives and gives up the same amount of heat (Q_1) it suffers no final change. But what we have just described is a process whose only final result is to transform into work the heat extracted from a source at the same temperature T_2 throughout. This is contrary to Kelvin's postulate.

• From the definition of efficiency it follows that

$$\eta = 1 + \frac{Q_2}{Q_1} = 1 - \frac{T_2}{T_1}$$

$$\frac{T_2}{T_1} = -\frac{Q_2}{Q_1} \quad (Q_1 > 0, Q_2 < 0).$$

Based on the above relationship an important application of Carnot cycle is that it offers a definition of the absolute temperature (Kelvin) scale based on pure thermodynamic arguments. For this we only need to have a value for a given temperature. An example for such value could be the melting point of ice. As such if we choose for the melting point of ice the value $T_1 = 273\,°C$ every other temperature T_2 can be defined via a Carnot cycle operating between heat source T_1 (consisting of melting ice) and a heat source T_2. If we then measure Q_1 and Q_2 we can apply the above relationship to calculate T_2. This thermodynamic definition of the Kelvin scale allows us to present a thermodynamic definition of absolute zero: the absolute zero is the temperature of a heat source (assumed to be the cold source of a Carnot cycle) which absorbs no heat (i.e. $Q_2 = 0$) even though the cycle produces work.

- In reality natural processes are irreversible processes. However, if some processes happen very slowly and the heat losses are small enough not to affect the temperature (which supposes that friction is also negligible), then the results will not differ from those expected from perfectly reversible processes. Such processes are often called partially reversible. If on the other hand friction and heat flow are important then the degree of reversibility decreases. In nature the degree of reversibility varies between almost perfectly reversible and perfectly irreversible.

  If in a thermal engine we increase the degree of irreversibility (for example, by increasing friction or direct leakages of heat), then the work from a given amount of heat will decrease. If the thermal losses keep on increasing we will end up with a perfectly irreversible engine. It follows that the efficiency of a thermal engine can take on values from zero (corresponding to the worst thermal engine operating at perfectly irreversible conditions) to some maximum value (corresponding to the best engine operating at perfectly reversible conditions). The Carnot cycle is a perfectly reversible cycle and as such it gives the maximum value the efficiency can take. This expresses the Carnot theorem which states that it is impossible to construct an engine operating between two heat sources which would have an efficiency greater than the efficiency of a Carnot cycle operating between the same heat sources. However, this maximum value will not be equal to one as this would require that $T_2 = 0$ K which is practically impossible. Recall from section 5.1 that

  $$Q_1 = nR^*T_1 \ln \frac{V_2}{V_1}$$

  and

  $$Q_2 = nR^*T_2 \ln \frac{V_4}{V_3}.$$

  It follows that

  $$\frac{Q_1}{T_1} = nR^* \ln \frac{V_2}{V_1}$$

and

$$\frac{Q_2}{T_2} = -nR^* \ln \frac{V_2}{V_1}$$

or

$$\frac{Q_1}{T_1} + \frac{Q_2}{T_2} = 0 \qquad (Q_1 > 0, Q_2 < 0). \qquad (5.5)$$

Any cyclic process can be divided into N (reversible) Carnot cycles, each one of them enclosed by two adiabats and two isotherms. If $N \longrightarrow \infty$ we can assume that the original cyclic process is well represented by the N Carnot cycles. In this case equation (5.5) can be extended to any cyclic process:

$$\frac{Q_1}{T_1} + \frac{Q_2}{T_2} + \frac{Q_3}{T_3} + \frac{Q_4}{T_4} + \cdots + \frac{Q_{2N-1}}{T_{2N-1}} + \frac{Q_{2N}}{T_{2N}} = 0$$

or

$$\sum_{i=1}^{2N} \frac{Q_i}{T_i} = 0.$$

In the limit where $N \longrightarrow \infty$ the above equation takes the form

$$\oint_{\text{rev}} \frac{\delta Q}{T} = 0 \qquad (5.6)$$

where rev stands for reversible processes. So with the aid of the Carnot cycle we may introduce a quantity that is an exact differential. This quantity, $\delta Q/T$, is defined as dS, where S is a new state function called *entropy*.

Now by considering equation (5.4) we can, for the reversible Carnot cycle, rewrite equation (5.5) as

$$\frac{Q_1 \eta_{\text{rev}}}{T_1 - T_2} + \frac{Q_2}{T_2} = 0.$$

Since Q_2/T_2 is less than zero it follows that $Q_1 \eta_{\text{rev}}/(T_1 - T_2)$ must be greater than zero. Since their sum must equal zero and, as we discussed above, the value of η is maximum for perfectly reversible processes, it follows that for irreversible processes where η becomes smaller

$$\frac{Q_1 \eta_{\text{irrev}}}{T_1 - T_2} + \frac{Q_2}{T_2} < 0.$$

The above arguments can be extended to any cyclic process represented as a sum of N subcycles to arrive at the general expression for any cycle

$$\oint \frac{\delta Q}{T} \leq 0$$

where the equality applies to reversible processes. Since $\delta Q/T$ is, for reversible processes, an exact differential it follows that for any reversible transformation $i \longrightarrow f$

$$\int_i^f \frac{\delta Q}{T} = \Delta S = S_f - S_i. \tag{5.7}$$

In other words the change in entropy depends only on the initial and final states, not on the particular transformation. This result is fundamental in the formulation of the second law.

Since S is a state function it follows that entropy changes are due to changes in both temperature and volume (or pressure). Using the first law we have that for a *reversible* process

$$dS = \frac{\delta Q}{T} = C_V \frac{dT}{T} + p \frac{dV}{T}$$

or

$$dS = C_V \frac{dT}{T} + nR^* \frac{dV}{V}$$

or

$$\frac{dS}{C_V} = \frac{dT}{T} + \frac{nR^*}{C_V} \frac{dV}{V}$$

or

$$\frac{dS}{C_V} = \frac{dT}{T} + (\gamma - 1) \frac{dV}{V}$$

or

$$S_f = S_i + C_V \ln \left(\frac{T_f V_f^{\gamma-1}}{T_i V_i^{\gamma-1}} \right). \tag{5.8}$$

Equation (5.8) provides the change in entropy of an ideal gas (since the ideal gas law was used) as a function of the initial and final temperature and volume. This relation cannot, therefore, be used for liquids or solids. For liquids or solids $C_V \equiv C_p = C$ and one can show (but we will not go into this proof here) that in this case the entropy change is dominated by the temperature change. As such for liquids or solids and for *reversible* processes

$$S_f - S_i = C \ln \frac{T_f}{T_i} \tag{5.9}$$

5.3 More on entropy

As in the case of energy, we will choose an arbritary equilibrium state O as the standard state where by definition S_O is zero. Then we will define the entropy S_A of an equilibrium state A as

$$S_A = \int_O^A \frac{\delta Q}{T}$$

where the integral is taken over a reversible transformation. It can easily be shown that if instead of a standard state O we had chosen

Figure 5.2
A cyclic transformation
consisting of a reversible
part from i to f (R) and
an irreversible part from
f to i (I).

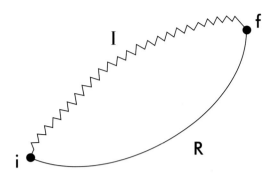

a standard state O' then the entropy of state A would differ from
the original one by an additive constant: from the above equation
we have

$$S'_A = \int_{O'}^{A} \frac{\delta Q}{T}.$$

From equation (5.7) it follows that

$$S'_A = S_A - S_{O'}$$

or

$$S_A - S'_A = S_{O'}.$$

Since O' is fixed, $S_{O'}$ is some constant. The entropy is thus defined except for an additive constant. As in the case of energy this
does not present a problem as long as we are dealing with differences, not actual values of entropy. Now let us consider the cyclic
transformation shown in figure 5.2. The path $i \xrightarrow{R} f$ represents a
reversible transformation and the path $f \xrightarrow{I} i$ represents an irreversible transformation. As we know, for any cycle $\oint \delta Q/T \leq 0$.
As such

$$\oint_{iRfIi} \frac{\delta Q}{T} \leq 0$$

or

$$\left[\int_{i}^{f} \frac{\delta Q}{T} \right]_R + \left[\int_{f}^{i} \frac{\delta Q}{T} \right]_I \leq 0.$$

Since $i \xrightarrow{R} f$ is reversible we have that

$$S_f - S_i = \left[\int_{i}^{f} \frac{\delta Q}{T} \right]_R.$$

Combining the last two equations leads to

$$S_f - S_i + \left[\int_{f}^{i} \frac{\delta Q}{T} \right]_I \leq 0$$

or

$$\left[\int_f^i \frac{\delta Q}{T}\right]_I \leq S_i - S_f.$$

By interchanging i and f we conclude that for any process

$$\int_i^f \frac{\delta Q}{T} \leq S_f - S_i \tag{5.10}$$

or

$$\Delta S \geq \int_i^f \frac{\delta Q}{T} \tag{5.11}$$

or

$$dS \geq \frac{\delta Q}{T} \tag{5.12}$$

where the equality applies to reversible processes only. Equation (5.12) is the general expression of the second law of thermodynamics and it indicates that the upper bound to the heat that can be absorbed by the system during a given change is $\delta Q = T dS$. For a completely isolated system ($\delta Q = 0$) equation (5.10) translates to

$$S_f \geq S_i. \tag{5.13}$$

This is an important conclusion as it indicates that for any spontaneous irreversible transformation (i.e. one not related to external influences) occurring in an isolated system, the final entropy is greater than the initial entropy. It follows that when an isolated system has attained a state of maximum entropy it cannot undergo any further transformation because any change would decrease its entropy, and this is not allowed by equation (5.13). Thus the state of maximum entropy is a state of stable equilibrium. Note that if the isolated system consists of a number of microscopic subsystems it would be possible for some of the subsystems to reduce their entropy but only at the expense of the rest of the subsystems, whose entropy must increase enough that the overall entropy of the system increases. A perfect isolated system is our universe (assuming that no other universes exist with which our universe can interact). It follows that the entropy of our universe increases in time. Equations (5.10)–(5.12) as well as equation (5.13) are the basic mathematical expressions of the second law of thermodynamics.

5.4 Special forms of the second law

Given the above definitions of the second law we can derive its expression for special cases such as the following.

• Finite isothermal transformations

From equation (5.11) it follows that in this case

$$\Delta S \geq \frac{1}{T} \int_i^f \delta Q$$

or

$$\Delta S \geq \frac{Q}{T}. \tag{5.14}$$

• Adiabatic transformations

Using equation (5.12) we obtain

$$dS \geq 0. \tag{5.15}$$

• Isentropic transformations

An isentropic transformation is one during which the entropy does not change. In this case it is clear that (recall equation (5.12))

$$\delta Q \leq 0. \tag{5.16}$$

Note that according to equation (5.15) a reversible adiabatic process is isentropic.

• Isochoric transformations

From the first law we have that when $dV = 0$ then $\delta Q = C_V dT$. As such for isochoric transformations

$$dS \geq C_V \frac{dT}{T}$$

or

$$\Delta S \geq C_V \ln \frac{T_f}{T_i}. \tag{5.17}$$

• Isobaric transformations

In this case

$$\delta Q = C_p dT$$

and it follows that

$$dS \geq C_p \frac{dT}{T}$$

or

$$\Delta S \geq C_p \ln \frac{T_f}{T_i}. \tag{5.18}$$

From equation (5.14) it follows that irreversible work increases a system's entropy. From equation (5.17) it follows that in the absence of work the change in entropy depends on the relation between T_f and T_i.

5.5 Combining the first and second laws

Consider the following form of the first law:

$$\delta Q = C_p dT - V dp.$$

Combining the above equation with $dS \geq \frac{\delta Q}{T}$ yields

$$T dS \geq C_p dT - V dp.$$

Recalling that $C_p = dH/dT$ we can reduce the above equation to

$$T dS \geq dH - V dp$$

or

$$dH \leq T dS + V dp. \tag{5.19}$$

Similarly, from

$$\delta Q = C_V dT + p dV$$

we arrive at

$$T dS \geq dU + \delta W$$

or

$$dU \leq T dS - p dV. \tag{5.20}$$

It is often convenient to introduce two new functions: the Helmholtz function $F = U - TS$ and the Gibbs function $G = H - TS = U + pV - TS$. Since $S = S(T, V)$ and $U = U(T)$ it follows that both F and G are state functions (i.e. $F = F(T, V), G = G(T, V)$), and thus are exact differentials. The advantage of these functions is that they can be used to express equations (5.19) and (5.20) in a form where the pairs (T, p) and (T, V) appear as the independent variables instead of the pairs (S, p) and (S, V). In this case equations (5.20) and (5.19) can be written as:

$$dF \leq -S dT - p dV \tag{5.21}$$

and

$$dG \leq -S dT + V dp. \tag{5.22}$$

The interpretation of the above functions is that for isothermal processes $dF \leq -\delta Q$ or $dF \leq -\delta W$ which makes F the energy available for conversion into work. The usefulness of G becomes clearer for isothermal-isobaric transformations. During such transformations (which apply in phase transitions, for example water to vapor) $dG = 0$ and therefore G is conserved. Relations (5.19)–(5.22) are often referred to as the fundamental relations.

5.6 Some consequences of the second law

Thermodynamic definition of temperature

As we know, the integral $\int_i^f \delta Q$ depends on the path from i to f, not just the states i and f. In this chapter we showed that the integral $\int_i^f \delta Q/T$ depends only on the initial and final states, not on the particular path from i to f. As such $1/T$ is the integration factor that makes δQ an exact differential. It follows that in ther-modynamical terms *temperature is the inverse of the integration factor of the differential of heat in reversible processes.*

The statistical nature of thermodynamics

The change in entropy during a reversible cooling at constant pres-sure of m grams of dry air from T_1 to T_2 $(T_1 > T_2)$ is

$$\Delta S = \int_{T_1}^{T_2} \frac{\delta Q}{T}.$$

Since $p = $ constant we have that $\delta Q = mc_{pd}dT$ where c_{pd} is the specific heat capacity of dry air. It follows that

$$\Delta S = \int_{T_1}^{T_2} mc_{pd}\frac{dT}{T}$$

or assuming that c_{pd} does not vary significantly with temperature

$$\Delta S = S_{T_2} - S_{T_1} = mc_{pd}\ln\frac{T_2}{T_1}.$$

Since m and c_{pd} are greater than zero and $\ln\frac{T_2}{T_1} < 0$ it follows that $\Delta S < 0$. Thus, the entropy decreases during the cooling. This de-crease does not violate the second law. Even though the dry air de-creases its entropy the environment that provided the heat suffers a positive change so that the overall entropy change of the system is positive. We can extend the above arguments to show that as the temperature goes down so does the entropy (or as temperature goes up so does the entropy). The question is why and what is the physical meaning of this. Kinetic theory provides an explanation for the above. According to the theory, at very low temperatures the motion of the molecules is very slow with the molecules being more or less uniformly distributed in space. This is a picture of a high order, with complete order (motionless molecules uniformly spaced) presumably achieved at absolute zero. As the temperature increases the motion of the molecules increases and the order is soon destroyed. If we combine this with the above result it follows that we can associate entropy with order or disorder. Accordingly, decreasing entropy implies increasing order and increasing entropy

Figure 5.3
The situation in (a)
represents greater order
and a lower number of
complexions than in (b)
where the number of
complexions and disorder
have increased.

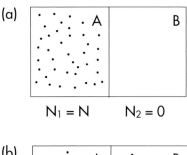

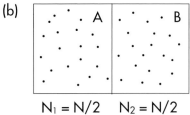

implies increasing disorder. This association justifies the word entropy, which in Greek means the "inner behavior" ($\epsilon\nu\tau\rho\sigma\pi\iota\alpha$). Now if such an association is correct we should be able to find a measure that defines disorder which should be mathematically related to entropy.

Let us consider a volume V consisting of two compartments A and B, with N total particles moving around. Let us denote with N_1 the number of particles in compartment A and with N_2 the number of particles in compartment B. Statistics tells us that the number of different ways, P, in which N_1 particles will be found in A and N_2 particles will be found in B is

$$P = \frac{N!}{N_1!N_2!}. \tag{5.23}$$

Now consider two cases as shown in figure 5.3. In the first case all particles are found in A (figure 5.3(a)), and in the second case the particles are all over the box (figure 5.3(b)). The first case ($N_1, = N, N_2 = 0$) obviously represents a state of more order than the second where it can be assumed that $N_1 = N_2 = N/2$. Applying equation (5.23) to both cases yields

1st case:

$$P = \frac{N!}{N!0!} = 1$$

2nd case:

$$P = \frac{N!}{N/2!\,N/2!} \gg 1.$$

We thus see that we have a measure P (often called the number of complexions) that increases as disorder increases. If we consider the first case as the initial state and the second case as the final state we can conclude (since $i \rightarrow f$ is an irreversible process) that as the entropy increases P increases. So P is a possible candidate to relate disorder to entropy. But how can we find their functional relationship?

The derivation of this relationship (one of the most beautiful yet simplest ones) is due to Boltzmann (Fermi (1936)). From the above we saw that as S increases, P increases and vice versa. So let us start with a general relationship of the form

$$S = f(P).$$

Next, consider a system consisting of two subsystems and and let S_1 and S_2 be the entropies and P_1 and P_2 the corresponding number of complexions. Then, $S_1 = f(P_1)$ and $S_2 = f(P_2)$. For the whole system we can then write that

$$S = S_1 + S_2$$

and

$$P = P_1 P_2$$

or

$$f(P_1 P_2) = f(P_1) + f(P_2)$$

i.e. the function f obeys the functional equation

$$f(xy) = f(x) + f(y).$$

Since the above equation is true for all values of x and y we may take $y = 1 + \epsilon$ with $\epsilon \ll 1$. Then we can write it as

$$f(x + x\epsilon) = f(x) + f(1 + \epsilon). \tag{5.24}$$

Expanding both sides of the above equation using Taylor's theorem and neglecting terms higher than first order we get

$$f(x) + x\epsilon f'(x) = f(x) + f(1) + \epsilon f'(1) \tag{5.25}$$

where f' indicates the first derivative. For $\epsilon = 0$ from equation (5.24) we have that $f(1) = 0$. Therefore, equation (5.25) reduces to

$$x\epsilon f'(x) = \epsilon f'(1)$$

or

$$xf'(x) = f'(1).$$

Since $f'(1)$ is a derivative evaluated at a value of one it is a constant. Thus

$$xf'(x) = k$$

or

$$f'(x) = \frac{k}{x}$$

or

$$f(x) = \int \frac{k}{x}\, dx$$

or

$$f(x) = k \ln x + \text{ constant}$$

or by changing $x \rightarrow P$,

$$S = k \ln P + \text{ constant.} \tag{5.26}$$

This equation relates order and entropy and constitutes a fundamental relationship connecting statistical mechanics and thermodynamics, thereby giving thermodynamics a statistical character.

Entropy and potential temperature

Recall that the potential temperature θ is given by

$$\theta = T \left(\frac{1000}{p} \right)^{R/c_p}.$$

By taking the logarithms of both sides we obtain

$$\ln \theta = \ln T + \frac{R}{c_p} \ln 1000 - \frac{R}{c_p} \ln p$$

or

$$c_p d \ln \theta = c_p d \ln T - R d \ln p. \tag{5.27}$$

From the first law we have

$$\delta Q = C_p dT - V dp$$

or

$$\frac{\delta Q}{T} = C_p \frac{dT}{T} - V \frac{dp}{T}$$

or

$$\frac{\delta Q}{T} = C_p \frac{dT}{T} - mR \frac{dp}{p}$$

or (for reversible processes)

$$ds = c_p \frac{dT}{T} - R \frac{dp}{p} \tag{5.28}$$

where ds is the specific entropy. Combining (5.28) and (5.27) leads to

$$ds = c_p d \ln \theta \tag{5.29}$$

or

$$s = c_p \ln \theta + \text{ constant.} \tag{5.30}$$

Thus, except for an additive constant, the specific entropy of a system is given by the logarithm of the potential temperature.

When θ remains constant the entropy remains constant. It follows that a reversible adiabatic process is isentropic. For an irreversible adiabatic processes $ds > 0, d\theta = 0$. In this case, the increase in entropy comes from irreversible work (for example, dissipation of kinetic energy due to friction). It follows that an isentropic process is adiabatic, but an adiabatic process may not be isentropic.

Maximization of entropy in the atmosphere

Recall that for isolated systems the state of maximum entropy corresponds to the most stable state of the system. The maximization of entropy results in some interesting insights about the stability of parcels or layers in the atmosphere, one of which is that for an isolated layer the potential temperature is constant with height (the derivation is beyond the scope of this book but it can be found in Bohren and Albrecht (1998)). In this case, assuming dry air conditions, we have from equation (5.27) that

$$\frac{1}{\theta}\frac{d\theta}{dz} = \frac{1}{T}\frac{dT}{dz} - \frac{R}{pc_{pd}}\frac{dp}{dz} = 0$$

or

$$\frac{dT}{dz} = \frac{RT}{pc_{pd}}\frac{dp}{dz}$$

or assuming hydrostatic conditions

$$\frac{dT}{dz} = \frac{RT}{pc_{pd}}(-\rho g)$$

or

$$\frac{dT}{dz} = -\frac{g}{c_{pd}}$$

which is the dry adiabatic lapse rate. Thus, maximization of entropy in the atmosphere implies that the equilibrium temperature of isolated layers decreases at the dry adiabatic rate (assuming no condensation or evaporation takes place). In other words the equilibrium temperature profile in a layer is *not* isothermal. This is quite the opposite of what will happen to an isolated solid that initially exhibits temperature gradients. In this case conduction will equalize the temperature gradients and uniformity will be achieved. That this is not the case in the atmosphere is explained by the fact that in the atmosphere the main mechanism of energy transfer is not conduction but convection. Convection causes mixing which results in increasing entropy (see solved example 5.1). Continuing mixing leads to maximization of entropy that results, as we derived above, in an equilibrium state where the temperature profile is that of the dry adiabatic lapse rate.

Since entropy and potential temperature are related it follows that potential temperature will also be related to stability. This

Figure 5.4
A hypothetical upper-level
flow and the trajectory
followed by a parcel.

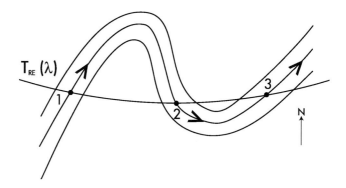

provides to meteorologists a way to characterize stability in the
atmosphere using a more familiar variable than entropy. More on
potential temperature and stability will be discussed in chapter 8
which is dedicated to stability in the atmosphere.

Atmospheric motions

Figure 5.4 shows a hypothetical upper-level flow close to some latitude λ. Along this latitude the radiative-equilibrium temperature
$T_{\mathrm{RE}}(\lambda)$ is assumed to be the same. Such thermal structure can be
achieved if the motion is completely zonal (i.e. parallel to latitude
circles), since air parcels then have infinite time to adjust to local
thermal equilibrium. If at position 1 a parcel is allowed to move
with the flow two possibilities exist. If the motion is very slow the
parcel has time to come into equilibrium with its surroundings, so
its temperature differs from T_{RE} only infinitesimally. Therefore, we
can assume that heat transfer along the trajectory of the parcel is
reversible. Then for positions 1 and 2 we have that

$$\int_1^2 c_p d\ln\theta = \int_1^2 ds = \int_1^2 \frac{\delta q}{T}.$$

If we assume that δq depends only on the parcel's temperature (for
example, $\delta q = Tf(T)$) then we can write the above equation as

$$\int_1^2 c_p d\ln\theta = \int_1^2 df(T) = f(T_2) - f(T_1).$$

Since $T_2 = T_1 = T_{\mathrm{RE}}$ the above relationship reduces to

$$\int_1^2 c_p d\ln\theta = 0$$

or

$$\theta_2 = \theta_1.$$

Thus, the parcel is restored to its initial thermodynamic state when
it returns to the original latitude. Since θ depends on p and p

depends on height it follows that under the slow motion conditions, successive crossings of the initial latitude result in zero net vertical motion and zero heat transfer.

On the other hand if the motion is fast enough that reversibility cannot be assumed, then

$$\int_1^2 c_p d\ln\theta < ds \neq 0$$

and therefore $\theta_2 \neq \theta_1$. In this case when the parcel returns to the initial latitude it is displaced vertically and the net heat transfer is not zero. Vertical motion resulting from irreversibility in the atmosphere plays an important role in maintaining meridional circulations, heat and moisture transfer etc.

Examples

(5.1) Imagine a volume V divided into two smaller volumes V_1 and V_2 by a partition. Assume that m_1 grams of an ideal gas at temperature T_1 occupy volume V_1 and m_2 grams of another ideal gas at temperature T_2 occupy volume V_2. Then the partition is removed. If the specific heat capacities at constant volume of the gases are c_{V_1} and c_{V_2} find (a) the limiting temperature of the mixture, and (b) the change in entropy for each gas and for the system as a whole.

(a) We assume that while the two gases exchange heat between them and perform work, the system as a whole does not exchange heat with its surroundings and its volume remains constant ($V = V_1 + V_2$). In other words for the whole system the process of mixing is adiabatic, with no work being done. As such, $\Delta U = 0$. The total internal energy before mixing is $C_{V_1}T_1 + C_{V_2}T_2$ and after the mixing it is $C_{V_1}T + C_{V_2}T$ (here we have made use of the approximation $U \approx C_V T+$ constant and then neglected the constant). Since $\Delta U = 0$ it follows that

$$C_{V_1}T_1 + C_{V_2}T_2 = C_{V_1}T + C_{V_2}T$$

which yields

$$T = \frac{C_{V_1}}{C_{V_1} + C_{V_2}}T_1 + \frac{C_{V_2}}{C_{V_1} + C_{V_2}}T_2. \qquad (5.31)$$

Considering that $C_{V_1} = c_{V_1}m_1$ and $C_{V_2} = c_{V_2}m_2$ the above equation becomes

$$T = \frac{c_{V_1}m_1}{c_{V_1}m_1 + c_{V_2}m_2}T_1 + \frac{c_{V_2}m_2}{c_{V_1}m_1 + c_{V_2}m_2}T_2$$

(b)   From equation (5.8) and the additive property of entropy it follows that for each gas the total entropy change is the sum of the entropy change due to the temperature change of the gas alone and the entropy change due to the volume change of the gas alone i.e.

$$\Delta S_1 = \Delta S_{1T} + \Delta S_{1V}$$
$$\Delta S_2 = \Delta S_{2T} + \Delta S_{2V}$$

where the lower limits for $\Delta S_{1T}, \Delta S_{1V}, \Delta S_{2T}$ and ΔS_{2V} are given by (recall equation 5.8)

$$\Delta S_{1T} = C_{V_1} \ln \frac{T}{T_1}$$

$$\Delta S_{2T} = C_{V_2} \ln \frac{T}{T_2}$$

$$\Delta S_{1V} = C_{V_1} \ln \left(\frac{V_1 + V_2}{V_1} \right)^{\gamma_1 - 1} = m_1 R_1 \ln \frac{V_1 + V_2}{V_1}$$

and

$$\Delta S_{2V} = C_{V_2} \ln \left(\frac{V_1 + V_2}{V_2} \right)^{\gamma_2 - 1} = m_2 R_2 \ln \frac{V_1 + V_2}{V_2}$$

where R_1 and R_2 are the specific gas constants of the two gases. For the system as a whole the lower limit for the total entropy change is

$$\Delta S = \Delta S_1 + \Delta S_2$$

or

$$\Delta S = (\Delta S_{1T} + \Delta S_{2T}) + (\Delta S_{1V} + \Delta S_{2V}) = \Delta S_T + \Delta S_V$$

or

$$\Delta S = \left[c_{V_1} m_1 \ln \frac{T}{T_1} + c_{V_2} m_2 \ln \frac{T}{T_2} \right]$$

(5.32)

$$+ \left[m_1 R_1 \ln \frac{V_1 + V_2}{V_1} + m_2 R_2 \ln \frac{V_1 + V_2}{V_2} \right].$$

In equation (5.32) the second term on the right hand side (ΔS_V) is necessarily greater than zero. The first term (ΔS_T) is either positive or zero when $T_1 = T_2$ (see problem 5.13). As such for the whole system $\Delta S > 0$ which according to (5.12) indicates irreversible process. This proves that mixing is irreversible.

(5.2)   Calculate the change in air pressure if the entropy decreases by 0.05 Jg^{-1} K^{-1} and the air temperature decreases by 5%.

Assuming that the process is reversible we have

$$dS = C_p \frac{dT}{T} - \frac{V}{T} dp$$

$$= C_p \frac{dT}{T} - nR^* \frac{dp}{p}.$$

Since we do not have any information on the mass of the air involved in the process we will write the above per unit mass:

$$ds = c_p \frac{dT}{T} - R \frac{dp}{p}$$

or

$$\frac{dp}{p} = \frac{c_p}{R} \frac{dT}{T} - \frac{ds}{R}$$

or considering the air as dry

$$\frac{dp}{p} = \frac{1005}{287}(-0.05) - \frac{(-50)}{287}$$

or

$$\frac{dp}{p} = -0.0009.$$

The pressure will decrease by a small amount of 0.09%.

(5.3)   In example 4.3, what is the change in entropy in each of the three transformations?

(a)   The first transformation is an isothermal reversible compression. Thus,

$$\Delta S = \int_i^f \frac{\delta Q}{T} = \frac{1}{T} \int_i^f \delta Q = \frac{Q}{T} = nR^* \ln \frac{V_f}{V_i} = nR^* \ln \frac{p_i}{p_f}.$$

(b)   The second transformation consists of two branches. The first one is a reversible adiabatic transformation. Thus, the entropy change in the first branch is zero. The second branch is an isobaric reversible compression. As such the entropy change of the second transformation is (recall that since T_i and T_f lie on the same isothermal $T_i = T_f = T$)

$$\Delta S = \int_m^f \frac{\delta Q}{T} = C_p \int_m^f \frac{dT}{T} = C_p \ln \frac{T_f}{T_m} = C_p \ln \frac{T_i}{T_m}$$

$$= C_p \ln \left(\frac{p_f}{p_i} \right)^{\frac{1-\gamma}{\gamma}}$$

$$= -nR^* \ln \frac{p_f}{p_i}$$

$$= nR^* \ln \frac{p_i}{p_f}$$

(c) The third transformation also consists of two branches:
one reversible isochoric and one reversible isobaric.
The total change in entropy is the sum of the changes
in entropy in the two branches

$$\Delta S = \Delta S_1 + \Delta S_2 = \int_i^{m'} \frac{\delta Q}{T} + \int_{m'}^f \frac{\delta Q}{T}$$

$$= \int_i^{m'} C_V \frac{dT}{T} + \int_{m'}^f C_p \frac{dT}{T}$$

$$= C_V \ln \frac{T_{m'}}{T_i} + C_p \ln \frac{T_f}{T_{m'}}$$

During this transformation $p_i V_i = nR^* T_i$ and $p_f V_i = nR^* T_{m'}$. Thus

$$T_{m'} = \frac{p_f}{p_i} T_i.$$

If we substitute the above value of $T_{m'}$ in the previous
equation we obtain

$$\Delta S = C_V \ln \frac{p_f}{p_i} + C_p \ln \frac{T_f}{T_i} \frac{p_i}{p_f}.$$

Since $T_f = T_i = T$ it follows that

$$\Delta S = C_V \ln \frac{p_f}{p_i} + C_p \ln \frac{p_i}{p_f}$$

or

$$\Delta S = nR^* \ln \frac{p_i}{p_f}.$$

Thus, in all transformations the change in entropy is the
same.

Problems

(5.1) Starting from the first law show that $\delta Q/T$ is an exact differential.

(5.2) Derive the so-called Maxwell relations for reversible processes

$$\left(\frac{\partial T}{\partial V} \right)_S = - \left(\frac{\partial p}{\partial S} \right)_V$$

$$\left(\frac{\partial S}{\partial V} \right)_T = \left(\frac{\partial p}{\partial T} \right)_V$$

$$\left(\frac{\partial T}{\partial p} \right)_S = \left(\frac{\partial V}{\partial S} \right)_p$$

$$\left(\frac{\partial S}{\partial p} \right)_T = - \left(\frac{\partial V}{\partial T} \right)_p.$$

(5.3) Find the lower limit for the change in specific entropy of dry air at a given temperature and pressure which is allowed to expand freely into an insulated chamber to assume twice its original volume. (200 J K^{-1} kg^{-1})

(5.4) What is the form of isenthalpic lines in a (p, V) diagram?

(5.5) Dry air at $T = 0\,^\circ$C and $p = 1$ atmosphere is compressed isentropically until its pressure becomes 10 atmospheres. Find the final temperature. (527.5 K)

(5.6) In a Carnot cycle the *area* for each transformation corresponds to the following work (a) isothermal expansion: $31\,165$ J, (b) adiabatic expansion: $21\,517$ J, (c) isothermal compression: $24\,282$ J, (d) adiabatic compression: $21\,517$ J. Calculate (1) the heat provided, per cycle, to the gas by the warm source, and (2) the efficiency of the cycle. (6883 J, 0.22)

(5.7) One mole of ideal gas occupies 5 liters at 273 K. If it expands into a vacuum until its volume becomes 20 liters, what is the lower limit for the change in entropy and the upper limit for the change in Gibb's function? (11.5 J K^{-1}, -3146 J)

(5.8) A fixed volume $2V$ is divided into two equal volumes V separated by a diaphragm. Each volume contains the same amount of the same gas at the same temperature T. Using Boltzmann's expression for entropy, show that after the diaphragm is removed the entropy change is zero. How will you approach this problem if the temperature in the two volumes is different? Use the approximation $\ln N! = N \ln N - N$.

(5.9) In a thermodynamic process a parcel of dry air is lifted in the atmosphere such that its pressure decreases from 1000 mbar to 800 mbar while its temperature remains constant. Calculate the change in the specific entropy of the parcel. (64 J kg^{-1} K^{-1})

(5.10) During a process a parcel of dry air descends from 900 to 950 mbar and its specific entropy decreases by 30 J kg^{-1} K^{-1}. If its initial temperature is 273 K what is (a) its final temperature and (b) its final potential temperature. (269 K, 273 K)

(5.11) Show that for flows on a constant pressure level the relative change (i.e. percent change of original value) in temperature is equal to the relative change in potential temperature.

(5.12) Cooler air moves over a warmer surface. What will happen to the surface pressure? (Hint: the condition of problem 5.11 does not apply here.)

(5.13) Show that the first term on the right hand side of equation (5.32) is greater than zero except when $T_1 = T_2$ when it is zero. (Hint: define $b = C_{V_1}/C_{V_2}$ and $x = T_2/T_1$ and find the first and second derivatives of $f(x) = \Delta S_T/C_{V_1}$.)

(5.14) Try to explain to your friends why the second law of thermodynamics implies a positive arrow of time (i.e. time goes only forwards).

(5.15) Consider two isolated systems existing very far from each other. The same irreversible transformation occurs in both of them. Will the change in entropy be the same in both of them?

(5.16) In an isentropic process the specific volume of a sample of dry air increases from $300 \text{ cm}^3 \text{ g}^{-1}$ to $500 \text{ cm}^3 \text{ g}^{-1}$. If the initial temperature was 300 K, what are the final values of T and p? (244.5 K, 1404 mbar)

(5.17) Prove geometrically (in a p, V diagram) equation (5.2).

CHAPTER SIX

Water and its transformations

The fundamental equations derived in the previous chapter can only be applied to closed systems (i.e. to systems that do not exchange mass) that are homogeneous (i.e. they involve just one phase). In such cases we do not need to specify how thermodynamic functions depend on the composition of the system. Only two independent variables (T and p, or p and V, or p and T) need to be known. Since the total mass (m) remains constant, if we know the values of the extensive variables per unit mass we can extend the equations to any mass by multiplying by m or by n (the number of moles).

A heterogeneous system involves more than a single phase. In this case we are concerned with the conditions of internal equilibrium between the phases. Even if the heterogeneous system is assumed to be (as a whole) a closed system the phases constitute homogeneous but open "subsystems" which can exchange mass between them. In this case the fundamental equations must be modified to include extra terms to account for the mass exchanges. These extra terms involve a function μ called the chemical potential, $\mu = \mu(T, p)$. We will not go into the details of defining μ; we will only accept that in the case of open systems something else must be included to account for the heterogeneity of the system. In this book we are concerned with a heterogeneous system that involves dry air (N_2, O_2, CO_2, Ar) and water, with the water existing in vapor and possibly one of the condensed phases (water or ice). Then one component of the system is the dry air (which is assumed to remain always unchanged and in a gaseous state) and the other component is the water substance which can exist in two phases. Thus dry air is a closed system whereas the two phases are two open systems. The system composed of dry air, water vapor, liquid water, and ice is also of interest but it is quite unstable and as a result it cannot exist in equilibrium.

6.1 Thermodynamic properties of water

In this chapter we will put aside dry air and we will concentrate on the one-component heterogeneous system "water" comprised of vapor and one of the condensed phases. Water vapor, like dry air, can be treated to a good approximation as an ideal gas. Accordingly, water vapor obeys the equation of state and if it existed only by itself its states would be determined by two independent variables. When, however, water vapor coexists with liquid water or ice things become a little complicated (or less complicated depending on the point of view). In this case, the mixture does not constitute an ideal gas and the equations applicable to ideal gases do not apply. One phase requires two independent variables (p and T) and the other phase two more (p' and T'). Then, equilibrium between the phases will require that $p = p'$ and $T = T'$. However, since the mass of each open subsystem does not remain constant another criterion for equilibrium must be considered. This criterion is given by $\mu = \mu'$ and expresses equilibrium in the terms responsible for the mass exchange. Accordingly, for equilibrium between the two phases the following three constraints must be satisfied:

$$p = p'$$
$$T = T'$$
$$\mu = \mu'.$$

The first two equations reduce the number of independent variables to two (say p and T). Then, the third equation $\mu = \mu(p, T) = \mu'(p', T') = \mu'(p, T)$ reduces the number of independent variables by one more. Since $\mu(p, T) = \mu'(p, T)$ it follows that $p = f(T)$. Thus, if we fix the temperature at which the phases exist in equilibrium the value of pressure also becomes fixed. This defines curves along which equilibrium between two phases can exist. We conclude that in the case of a one-component system involving two phases in equilibrium the number of independent variables (otherwise known as degrees of freedom) is one, not four.

Let us now consider the case of a one-component system involving all its phases (vapor, liquid, solid). Following the same arguments as above the following constraints must be satisfied for equilibrium

$$p = p' = p''$$

$$T = T' = T''$$

$$\mu = \mu' = \mu''.$$

Then the first two equations reduce the number of independent variables from six to two. The first equality of the third equation reduces them by one more and the second equality by yet one

Figure 6.1
Phase-transition equilibria
for water.

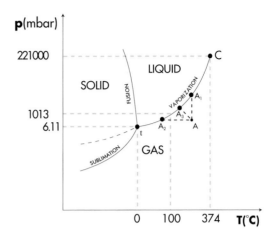

more. Thus, we remain with zero independent variables, meaning that all values are fixed, which implies that coexistence of all phases at equilibrium requires just a point, called the triple point, in the p, V, T state space. It follows that the greater the number of phases the fewer the degrees of freedom. The following formula suggested by Gibbs generalizes the above as it gives the number of independent state variables, N, for a heterogeneous system involving C different and non-reactive components in a total of P phases:

$$N = C + 2 - P. \tag{6.1}$$

Figure 6.1 shows the triple point for water as well as the curves $p = f(T)$ representing the phase transition equilibria. We call the three equilibria vaporization (gas $\leftrightarrow$ liquid), fusion (liquid $\leftrightarrow$ ice), and sublimation (gas $\leftrightarrow$ solid). Along the curves for vaporization and sublimation, vapor is in equilibrium with water and ice, respectively. As such these curves provide the equilibrium vapor pressure for water and ice. Vapor in equilibrium with a condensed phase is often called saturated and the corresponding equilibrium vapor pressure is called saturation vapor pressure. There is no real distinction between the two terms. Note that the extension of the vaporization curve to temperatures below the triple point corresponds to supercooled water, with vapor pressure over supercooled water being greater than that over ice. This is a metastable equilibrium where supercooled water and water vapor coexist. Here the system of supercooled water and water vapor may be stable with respect to small changes in temperature and pressure but introduction of ice in the system makes it unstable and as a result water freezes. The triple point corresponds to $p_t = 6.11$ mbar and $T = 273$ K (more accurately 273.16 K). Because of the different densities of liquid water, ice, and water vapor, the specific volumes for liquid water, a_w, ice, a_i, and water vapor, a_v, at the triple point are:

$$a_{\mathrm{w}} = 1.000 \times 10^{-3} \ \mathrm{m^3\,kg^{-1}}$$
$$a_{\mathrm{i}} = 1.091 \times 10^{-3} \ \mathrm{m^3\,kg^{-1}}$$
$$a_{\mathrm{v}} = 206 \ \mathrm{m^3\,kg^{-1}}.$$

An interesting feature of figure 6.1 is that the vaporization curve ends at point C where the temperature (T_C) is 374 °C and the pressure (p_C) is approximately 2.21×10^5 mbar. Beyond this critical point there is no line to separate liquid from vapor. Otherwise stated, beyond point C there is no discontinuity between the liquid and the gaseous phase. This indicates that beyond point C we cannot differentiate between the liquid and the vapor phase. In order to understand the significance of critical point C let us consider point A representing the temperature and the pressure of a sample of water vapor below critical point C. If we decrease the volume while keeping the temperature constant the pressure increases until it becomes equal to the pressure at point A_1 where it will condense. Similarly, if we keep the pressure constant and cool the gas until its temperature becomes equal to the temperature corresponding to point A_2, the gas will condense. Obviously the gas will condense if we perform any intermediate process described by line AA_3. Thus, there are three ways by which the gas can become liquid. We either have to cool it at constant pressure or compress it at constant temperature or simultaneously compress it and cool it. All three ways require crossing the equilibrium line. The fact that there is an end to the vaporization curve means that above T_C we cannot liquefy a gas by compressing it at a constant temperature and that above p_C we cannot liquefy a gas by cooling it at a constant pressure. Because in our atmosphere temperature and pressure are well below the critical point, water vapor can condense. For the other gases in the atmosphere (N_2, O_2, Ar) the critical temperatures are very low (-147 °C, -119 °C, -122 °C, respectively) and so they do not condense.

Let us now consider phase changes along isotherms in the (p, V) domain. Such a domain (shown in figure 6.2) is called an Amagat–Andrews diagram. We start with a sample of water vapor at a state corresponding to point A in figure 6.1 (i.e. at a temperature T_1 and pressure p_1 greater than the triple point temperature T_{t} and pressure p_{t}. If we compress the vapor isothermally the pressure increases until we reach point A_1. At this point liquid water and water vapor coexist in equilibrium. This means that some water vapor has condensed to form liquid water and that liquid and vapor are in equilibrium. Now what will happen if we further compress the vapor? Since we are on the equilibrium curve the pressure of the vapor depends only on temperature. As such, since the transformation is isothermal, as long as liquid and vapor coexist a

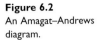

Figure 6.2
An Amagat–Andrews
diagram.

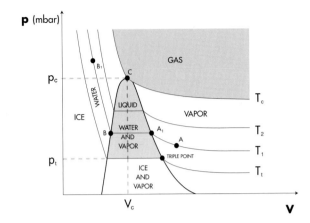

further decrease in volume cannot change the pressure. Do not get confused here. The ideal gas law *does not* apply to the mixture. Thus, if we keep on compressing the vapor we will observe that more vapor condenses to form liquid water (with liquid and vapor being always in equilibrium) until we reach point B where all vapor is condensed. This part will be represented in the (p, V) diagram by the horizontal (constant p) stretch $A_1 B$. A further (small) compression causes the liquid water's pressure to increase rapidly (owing to the low compressibility of liquids). This part is indicated by the segment BB_1. The complete isotherm is depicted in figure 6.2 by the line $AA_1 BB_1$.

For higher temperatures we observe that the horizontal stretch of the isotherms decreases until it reduces to a point. This is the point C corresponding to the critical state T_C, p_C, V_C. The isotherms above the critical temperature are monotonic decreasing functions with no discontinuity. They go over into equilateral hyperbolas which is how ideal gases behave. For temperatures below T_1 the horizontal stretch of the isotherms increases. The dashed line in figure 6.2 connects the beginning and end points of the horizontal stretches. This line together with the critical isotherm and the isotherm corresponding to the triple point temperature (T_t) partition the diagram into six regions: vapor, gas, liquid water, ice, liquid water plus vapor, and ice plus vapor. Here the distinction between gas and vapor reflects only the fact that above T_C no two-phase discontinuity for condensation exists.

6.2 Equilibrium phase transformations – latent heat

Before we proceed and to avoid confusion with too many subscripts we need to make a change in our notation. From now on we shall represent the water vapor pressure as e. Furthermore, during phase transitions the equilibrium (saturation) vapor pressure over water will be denoted as e_{sw} and equilibrium vapor pressure over ice will be denoted as e_{si}.

In the case of a *homogeneous* system undergoing an isobaric transformation that does not result in a phase change the heat exchanged is proportional to the temperature change (the proportionality constant is C_p). In the case of a *heterogeneous* system involving two phases we say that when the two phases are in equilibrium a fixed temperature implies a fixed pressure, meaning that isobaric transformations are also isothermal transformations. Thus, even though the masses of the two phases change, the temperature of the system does not (recall the horizontal stretches in figure 6.2). In such cases, the amount of heat exchanged depends only on mass changes (which modify the internal energy of each subsystem) and on the work done owing to possible volume changes (the volume can change because the mixture does not obey the ideal gas law). By definition, the latent heat, L, of a transformation is the heat absorbed (or given away) by the system during an isobaric phase transition

$$L = \delta Q_{p=\text{constant}}.$$

Since enthalpy is defined as $H = U + pV$, the above equation and the first law combine to yield

$$L = dH. \tag{6.2}$$

It follows that the latent heat of a phase change is the change in enthalpy during the transformation. The latent heats of vaporization (liquid $\leftrightarrow$ vapor), fusion (ice $\leftrightarrow$ liquid) and sublimation (ice $\leftrightarrow$ vapor) are denoted by L_v, L_f, and L_s respectively. Note that the above latent heats are positive when during the transformation heat is absorbed (liquid $\rightarrow$ vapor, ice $\rightarrow$ liquid, ice $\rightarrow$ vapor) and negative when heat is released (vapor $\rightarrow$ liquid, liquid $\rightarrow$ ice, vapor $\rightarrow$ ice). L_v, L_f, and L_s are also called the enthalpy of vaporization, fusion and sublimation, respectively. According to the above,

$$L_v = H_v - H_w = U_v - U_w + p_{wv}(V_v - V_w)$$
$$L_f = H_w - H_i = U_w - U_i + p_{wi}(V_w - V_i) \tag{6.3}$$
$$L_s = H_v - H_i = U_v - U_i + p_{vi}(V_v - V_i)$$

where here p stands for the equilibrium pressure between two phases. At the triple point $p_{wv} = p_{vi} = p_{wi} = 6.11$ mbar. It follows that at the triple point the specific latent heats l_v, l_f, l_s satisfy the relationship $l_s = l_f + l_v$. Obviously, the latent heats depend on temperature (see Table A3). Nevertheless, in the range of temperatures observed in the troposphere they do not vary significantly. Because of that it is often assumed that the latent heats

Figure 6.3
The isotherm T_1 of figure
6.2. For any given change
$C \longrightarrow C'$ the change in
volume is dV. This causes
an amount dm of liquid
water to evaporate.

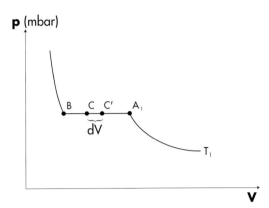

are independent of temperature. It is easy to show that the first
law in the case of an isobaric change of phase can be written as
$dU = L - pdV$. For fusion $dV \approx 0$ and for vaporization and sub-
limation $dV = V_{\text{vapor}} - V_{\text{liquid water or ice}} \approx V_{\text{vapor}}$. Thus, the first
law reduces to

$$\begin{aligned}
dU &= L & \text{for fusion} \\
dU &= L - mR_{\text{v}}T & \text{for vaporization or sublimation}
\end{aligned} \tag{6.4}$$

where R_{v} is the gas constant for water vapor.

6.3 The Clausius–Clapeyron (C–C) equation

For an ideal gas we saw how the equation of state $p = f(V, T)$
relates changes of pressure, temperature, and volume. In the case
of a heterogeneous system involving two phases we saw that $p =
e_{\text{s}} = f(T)$. Does this indicate that a relationship similar to the
equation of state for ideal gases exists for heterogeneous systems?

Let us consider the transformation described by isothermal T_1
in figure 6.2 and focus our attention on segment A_1B. Going from
A_1 to B a change of phase takes place (vapor $\rightarrow$ liquid). As we
discussed before, between A_1 and B vapor is in equilibrium with
liquid water and $p = e_{\text{sw}} = f(T)$. Because of that any other state
variable or function is a function of temperature only. Let us assume that at point C (figure 6.3) the mass of the liquid phase is
m_{w} and the mass of the vapor phase is m_{v}. Then the total volume
and internal energy of the system at point C are

$$V = m_{\text{w}}a_{\text{w}} + m_{\text{v}}a_{\text{v}}$$
$$U = m_{\text{w}}u_{\text{w}} + m_{\text{v}}u_{\text{v}}$$

where $a_{\text{w}}, a_{\text{v}}, u_{\text{w}}, u_{\text{v}}$ are the specific volumes and internal energies
of the two phases. Now assume that the system moves from C to
C'. This change corresponds to a change in volume dV and causes

an amount dm of liquid to evaporate. The volume at point C' will be

$$V + dV = (m_\mathrm{w} - dm)a_\mathrm{w} + (m_\mathrm{v} + dm)a_\mathrm{v}$$

or using the previous equation

$$dV = (a_\mathrm{v} - a_\mathrm{w})dm. \qquad (6.5)$$

Similarly

$$dU = (u_\mathrm{v} - u_\mathrm{w})dm. \qquad (6.6)$$

Recall from equation (6.3) that during this change

$$u_\mathrm{v} - u_\mathrm{w} + e_\mathrm{sw}(a_\mathrm{v} - a_\mathrm{w}) = l_\mathrm{v}. \qquad (6.7)$$

Dividing equation (6.6) by (6.5) yields

$$\frac{dU}{dV} = \frac{u_\mathrm{v} - u_\mathrm{w}}{a_\mathrm{v} - a_\mathrm{w}}$$

or using equation (6.7)

$$\frac{dU}{dV} = \frac{l_\mathrm{v}}{a_\mathrm{v} - a_\mathrm{w}} - e_\mathrm{sw}. \qquad (6.8)$$

Here is as a good place as any to restate that while general definitions such as the first law (equation (4.5)), the second law (equation (5.12)), enthalpy ($H = U + pV$), specific heat (equation (4.14)), etc. are valid for any system, variations or expressions derived using the ideal gas law (for example, equations (4.22), (4.18)) *are not* valid for any system but ideal gases. According to problem 6.1, for any system

$$\left(\frac{\partial U}{\partial V}\right)_T = T\left(\frac{\partial p}{\partial T}\right)_V - p. \qquad (6.9)$$

Also, according to problem 6.2, *for an ideal gas* $(\partial U/\partial V)_T = 0$ (Joule's law). However, water in equilibrium with its vapor is not an ideal gas and U and e_sw are functions of temperature only. Then since the transformation is isothermal we can write that

$$\left(\frac{\partial U}{\partial V}\right)_T = \frac{dU}{dV}$$

and

$$\left(\frac{\partial e_\mathrm{sw}}{\partial T}\right)_V = \frac{de_\mathrm{sw}}{dT}.$$

Accordingly, equation (6.9) becomes

$$\frac{dU}{dV} = T\frac{de_\mathrm{sw}}{dT} - e_\mathrm{sw}.$$

Combining this equation with equation (6.8) yields

$$T\frac{de_\mathrm{sw}}{dT} - e_\mathrm{sw} = \frac{l_\mathrm{v}}{a_\mathrm{v} - a_\mathrm{w}} - e_\mathrm{sw}$$

or

$$\frac{de_{\text{sw}}}{dT} = \frac{l_{\text{v}}}{T(a_{\text{v}} - a_{\text{w}})}. \tag{6.10}$$

Equation (6.10) is called the Clausius–Clapeyron (C–C) equation. In its general form,

$$\frac{de_{\text{s}}}{dT} = \frac{l}{T\Delta a}, \tag{6.11}$$

it relates the equilibrium pressure between two phases to the temperature of the heterogeneous system. Here l is the specific latent heat (or specific enthalpy) of the change of phase and Δa is the difference in the specific volumes between the two phases at temperature T. It is for heterogeneous systems what the equation of state is for ideal gases.

6.4 Approximations and consequences of the C–C equation

● Temperature dependence of enthalpy of vaporization

By differentiating equation (6.3) we have

$$\frac{\partial L_{\text{v}}}{\partial T} = \frac{\partial H_{\text{v}}}{\partial T} - \frac{\partial H_{\text{w}}}{\partial T} = C_{pv} - C_{pw}$$

where C_{pv} and C_{pw} are the heat capacities at constant pressure for vapor and liquid water, respectively. Since L_{v} is the enthalpy difference between the vapor and liquid water phases at equilibrium, it depends only on temperature. As such the above equation can be written as

$$\frac{dL_{\text{v}}}{dT} = C_{pv} - C_{pw}. \tag{6.12}$$

For a wide range of temperatures ($-20\,°\text{C}$ to $30\,°\text{C}$), C_{pv} and C_{pw} vary very little (about 1%). Because of that we can consider them independent of temperature and integrate equation (6.12) to get

$$L_{\text{v}} = L_{\text{v0}} + (C_{pv} - C_{pw})(T - T_0)$$

or

$$l_{\text{v}} = l_{\text{v0}} + (c_{pv} - c_{pw})(T - T_0) \tag{6.13}$$

where l_{v0} is the specific latent heat of vaporization at the reference state T_0. For the reference state $T_0 = 273\,\text{K}$ with $l_{\text{v0}} = 2.5 \times 10^6$ J kg^{-1}, $c_{pv} = 1850$ J kg^{-1}K^{-1}, and $c_{pw} = 4218$ J kg^{-1}K^{-1}, equation (6.13) provides a very good approximation of l_{v} over the range from $-20\,°\text{C}$ to $30\,°\text{C}$.

We can repeat the above for sublimation and fusion to arrive at

$$\frac{dL_s}{dT} = C_{pv} - C_{pi}$$

$$\frac{dL_f}{dT} = C_{pw} - C_{pi}$$

or

$$l_s = l_{s0} + (c_{pv} - c_{pi})(T - T_0) \tag{6.14}$$
$$l_f = l_{f0} + (c_{pw} - c_{pi})(T - T_0).$$

Recall that the specific heat for constant volume is $c_V = \delta q/dT$. Since water vapor is considered an ideal gas, it follows that

$$c_{Vv} = \frac{\delta q}{dT} = \frac{du_v}{dT}.$$

For liquid water

$$c_{Vw} = \frac{\delta q}{dT} = \frac{du_w}{dT} + p\frac{da_w}{dT}.$$

Since a_w varies very little with T the above equation reduces to

$$c_{Vw} = \frac{du_w}{dT}.$$

Similarly, for ice $c_{Vi} = du_i/dT$. Now since $h = u + pa$ it follows that for water vapor $c_{pv} = dh_v/dT = du_v/dT + R_v$. Therefore for water vapor $c_{pv} \neq c_{Vv}$. Liquid water and ice do not behave as ideal gases. Therefore,

$$c_{pw} = \frac{du_w}{dT} + p\frac{da_w}{dT} + a_w\frac{dp}{dT} \approx \frac{du_w}{dT}$$

(because $dp = 0$ and $da_w/dT \approx 0$). Thus, $c_{pw} \approx c_{Vw} = c_w$. Similarly $c_{pi} \approx c_{Vi} = c_i$. Using the values of $c_{pv}, c_{Vv}, c_w,$ and c_i at $T = 0\,^\circ\text{C}$ (1850, 1390, 4218, and 2106 J kg^{-1}K^{-1}, respectively) and using equations (6.13) and (6.14) we find that

$$\frac{dl_v}{dT} \approx -2368 \text{ J kg}^{-1}\text{K}^{-1}$$

$$\frac{dl_s}{dT} \approx -256 \text{ J kg}^{-1}\text{K}^{-1}$$

$$\frac{dl_f}{dT} \approx 2112 \text{ J kg}^{-1}\text{K}^{-1}.$$

These rates are small compared with the values of l_v, l_s, l_f at 0 °C (2.5×10^6, 2.834×10^6, 0.334×10^6 J kg^{-1}K^{-1}, respectively). We can thus conclude that all latent heats vary very little with temperature. As such the above equations justify approximating $l_v, l_s,$ and l_f as constants. An improvement to this approximation is given by equations (6.13) and (6.14) which express them as linear functions of T.

● Temperature dependence of equilibrium (saturation) vapor pressure

In the case of vaporization $a_v \gg a_w$ and C–C equation can be approximated as

$$\frac{de_{sw}}{dT} = \frac{l_v}{Ta_v}. \tag{6.15}$$

By combining equations (6.15) and (6.13) and considering vapor as an ideal gas (i.e. $e_{sw}a_v = R_vT$) we obtain

$$\frac{1}{e_{sw}}\frac{de_{sw}}{dT} = \frac{l_{v0} + (c_{pw} - c_{pv})T_0}{R_vT^2} - \frac{c_{pw} - c_{pv}}{R_vT}$$

or

$$\int_{e_{sw} \text{ at } T_0}^{e_{sw} \text{ at } T} \frac{de_{sw}}{e_{sw}} = \frac{l_{v0} + (c_{pw} - c_{pv})T_0}{R_v}\int_{T_0}^{T}\frac{dT}{T^2} - \frac{c_{pw} - c_{pv}}{R_v}\int_{T_0}^{T}\frac{dT}{T}$$

or

$$\ln\frac{e_{sw}}{e_{s0}} = \frac{l_{v0} + (c_{pw} - c_{pv})T_0}{R_v}\left(\frac{1}{T_0} - \frac{1}{T}\right) - \frac{c_{pw} - c_{pv}}{R_v}\ln\frac{T}{T_0}.$$

Considering as the reference state the triple point where $T_0 = 0\,°\text{C}$, $e_{s0} = 6.11$ mbar, $l_{v0} = 2.5 \times 10^6$ J kg^{-1}, $c_{pv} = 1850$ J kg^{-1} K^{-1}, and $c_{pw} = 4218$ J kg^{-1}K^{-1} the above equation reduces to

$$\ln\frac{e_{sw}}{e_{s0}} = 6808\left(\frac{1}{T_0} - \frac{1}{T}\right) - 5.09\ln\frac{T}{T_0}$$

or

$$e_{sw} = 6.11\exp\left(53.49 - \frac{6808}{T} - 5.09\ln T\right). \tag{6.16}$$

The above equation provides the relation between e_{sw} and T. Another way to establish such a relation would be to integrate equation (6.15) assuming that l_v is constant and independent of T. This would result in an approximation to the above equation given by

$$\ln\frac{e_{sw}}{e_{s0}} = \frac{l_v}{R_vT_0} - \frac{l_v}{R_vT}.$$

or

$$e_{sw} = 6.11\exp\left(19.83 - \frac{5417}{T}\right). \tag{6.17}$$

In both equations (6.16) and (6.17) e_{sw} is in mbar and T is in degrees Kelvin. In problem 6.3 you are asked to show that equations (6.16) and (6.17) are nearly identical for temperatures in the range $-20\,°\text{C}$ to $30\,°\text{C}$.

The corresponding relation to equations (6.16) and (6.17) for sublimation can be derived following the same procedure (here again $a_v \gg a_i$) with $l_s = 2.834 \times 10^6$ J kg^{-1} and $c_i = 2106$ J kg^{-1} K^{-1}. They are:

$$e_{si} = 6.11 \exp\left(26.16 - \frac{6293}{T} - 0.555 \ln T\right) \qquad (6.18)$$

and

$$e_{si} = 6.11 \exp\left(22.49 - \frac{6142}{T}\right). \qquad (6.19)$$

The C–C equation and its simplified versions describe the thermodynamic state of water vapor and liquid water (or water vapor and ice) when the two phases are in equilibrium. When the system's pressure is greater than the equilibrium pressure, equilibrium is approached by condensation (sublimation) of water vapor. This reduces the amount of vapor in the system thereby reducing the vapor pressure. If the system's pressure is lower than the equilibrium pressure, equilibrium is approached by water (ice) evaporating (sublimating). Because of the particular exponential dependence of e_{sw} and e_{si} on T in equations (6.17) and (6.19) it follows that for higher temperatures e_{sw} is greater than for lower temperatures. As such more water vapor can exist at higher temperatures. For example, for $T = 300$ K from equation (6.17) we have $e_{sw} = 36$ mbar. For $T = 273$ °C, $e_{sw} = 6$ mbar. This is a substantial difference. Note that this does necessarily imply that warmer air can hold more water vapor than colder air as it is often wrongly stated. All the above arguments and formulas were derived in the absence of air! We only considered two phases with water evaporating (or vapor condensing) into a vacuum. If you think about it this is consistent with Dalton's law which would indicate that in a mixture of air and water vapor the partial pressures are independent of the mixture. Therefore, keep in mind that when we say that "air at a given temperature is saturated" we do not imply that the air holds as much vapor as it can hold. We strictly mean that at that temperature the amount of vapor that can exist is maximum regardless of the presence of air.

The reason for the relation between equilibrium vapor pressure and temperature arises from purely kinetic reasons. At equilibrium evaporation is zero (strictly speaking it is the net evaporation = evaporation minus condensation that is zero as molecules inside water have as much chance to evaporate as vapor molecules have to condense). At higher temperatures the molecules inside the liquid acquire greater speeds and their chance to escape increases. As a result (net) evaporation increases.

The exponential dependence of e_{sw} (or e_{si}) on T has important consequences for our climate system. Tropical warm water can transfer more water vapor into the atmosphere than colder extratropical water. Most of this water vapor is then precipitated as a result of organized convection in the tropics thereby releasing large amounts of heat. This heat can then be converted to work and generate kinetic energy which helps maintain the general circulation against frictional dissipation.

• Changes in the melting and boiling points for water

At boiling point $(T = 100\ °C)$, $l_v = 2.26 \times 10^6$ J kg^{-1}, $a_v = 1.673$ m^3 kg^{-1}, and $a_w = 0.00104$ m^3 kg^{-1}. Then from equation (6.15) it follows that

$$\frac{de_{sw}}{dT} = 3621 \text{ Pa K}^{-1}$$

or

$$\frac{de_{sw}}{dT} = 0.03575 \text{ atm K}^{-1}.$$

Accordingly, a decrease in pressure by 0.03575 atm (36.2 mbar) lowers the boiling point by $1\,°C$. Note that if we evaluate the approximate relationship (6.16) at $T_1 = 373$ K and $T_2 = 372$ K, we will find that $e_{sw}(T_1) \approx 1003$ mbar and $e_{sw}(T_2) \approx 969$ mbar, or $\Delta e_{sw} \approx -34$ mbar. If we use instead equation (6.17), we will find that $e_{sw}(T_1) \approx 1237$ mbar, $e_{sw}(T_2) \approx 1190$ mbar, and $\Delta e_{sw} \approx -47$ mbar. The differences are due to the fact that in deriving the approximate expressions we have assumed that the specific heats and latent heats are either constant or linear functions of T. While in the range of temperature relevant to weather these approximations may be good, for temperatures outside this range the differences may be significant.

At melting point $(T = 0\ °C)$, $l_f = 0.334 \times 10^6$ J kg^{-1}, $a_w = 1.00013 \times 10^{-3}$ m^3 kg^{-1}, and $a_i = 1.0907 \times 10^{-3}$ m^3 kg^{-1}. Substituting these values into equation (6.15) yields

$$\frac{dp_{wi}}{dT} = -13\,503\,800 \text{ Pa K}^{-1}$$

or

$$\frac{dp_{wi}}{dT} = -133.3 \text{ atm K}^{-1}.$$

This result indicates that lowering the melting point by just one degree will require a tremendous increase in pressure. This is actually rather anomalous as in most cases the melting point increases with increasing pressure. The strange behavior of water and ice is due to the fact that ice is less dense than water whereas in most cases the solid is denser than the liquid. The fact that the melting point of ice is lowered by increasing pressure is very important to geophysics as it explains the motion of glaciers. When a mass of ice encounters a rock on the glacier bed the high pressure of the ice against the rock lowers the melting point of ice causing the ice to melt and then to flow around the rock. Once behind the rock the pressure is restored and ice freezes again. In this way ice propagates around obstacles.

Examples

(6.1)   Calculate the work done, the amount of heat absorbed, and the change in the internal energy during (a) melting of 1 gram of ice to water at a temperature of 0 °C and under a constant pressure of 1 atmosphere, and (b) evaporation of 1 gram of water to water vapor at a temperature of 100 °C and under a constant pressure of 1 atmosphere.

(a) When the effect of absorbing an amount Q at a constant pressure is a change in the physical state of the system (which also occurs at constant temperature), then Q is proportional to the mass that undergoes the change. In our case $Q = ml_f$ where l_f is the specific latent heat of fusion of 0 °C. For ice $l_f = 79.7$ cal g^{-1}. Thus,

$$Q = l_f m = 79.7 \text{ cal.}$$

During melting the volume changes. The corresponding work done is

$$W = \int_{ice}^{water} p dV = p \int_{ice}^{water} dV = p(V_{water} - V_{ice}).$$

The density of ice at 0 °C is 0.917 g cm^{-3}, so one gram occupies a volume of 1.09 cm^3. The density of water at 0 °C is 1 g cm^{-3}. Accordingly, one gram of water occupies a volume of 1.0 cm^3. It follows that

$$W = 1013 \times 10^2 \times (1 - 1.09) \times 10^{-6} \text{ J}$$

or

$$W = -0.00912 \text{ J}$$

or

$$W = -0.0022 \text{ cal.}$$

The work done is negative because the volume of the resulting water is smaller than the volume of ice. From the above we then find that

$$\Delta U = Q - W = 79.7022 \text{ cal.}$$

(b) In this case at 100 °C, $l_v = 539$ cal g^{-1}. Thus,

$$Q = l_v m = 539 \text{ cal.}$$

Water at 100 °C has a density of 0.958 g cm^{-3} which for one gram gives a volume of 1.04 cm^3. The volume of 1 gram of water vapor at 100 °C is 1673 cm^3. As such

$$W = p(V_{\text{vapor}} - V_{\text{water}}) = 1013 \times 10^2 \times (1673 - 1.04) \times 10^{-6}$$
$$= 169.37 \text{ J} = 40.47 \text{ cal}.$$

It then follows that

$$\Delta U = 498.53 \text{ cal}.$$

Note that in both cases the internal energy increases even though the temperature remains constant. This is due to the fact that the degrees of freedom increase. The degrees of freedom of the molecules in H_2O is much greater in water than ice and much greater in water vapor than water.

(6.2) **Can raindrops grow from cloud droplets through condensation alone?**

In order to address this question we have to make some reasonable assumptions. Cloud droplets form over nuclei and begin to grow according to the available water vapor. If the vapor pressure is greater than the equilibrium vapor pressure droplets grow by condensation. So we need to get an idea of how much vapor is available and how many droplets are going to compete for this vapor. Let us assume that the concentration of nuclei is 100/cm^3 which corresponds to rather clean air. Now we need to estimate the amount of vapor that will be available during the process of growing by condensation. We can get a gross estimate of this amount by assuming that at the level where clouds form (say 1000 m) the air is saturated and the surface temperature is about 30 °C. If this air ascends adiabatically, the temperature decreases. As the temperature decreases the equilibrium (saturation) vapor pressure decreases. This means that not as much vapor is needed for equilibrium or that the vapor pressure becomes greater than the equilibrium vapor pressure. Thus, the droplets begin to grow. If the air ascends to the maximum level (say the top of the troposphere at about 16 km), then we can estimate the equilibrium vapor pressure at that level by assuming that no water vapor condenses using the equation

$$\ln \frac{e_{\text{sw}}}{6.11} = 19.83 - \frac{5417}{T},$$

where $T = 30 - (16 \times 9.8) = -126.8$ °C $= 146.2$ K. It follows that at this level $e_{\text{sw}} = 0$. At cloud base where the temperature is ≈ 20.2 °C, $e_{\text{sw}} \approx 23.7$ mbar. As such in this scenario the equivalent of 23.7 mbar in water vapor will be available to the droplets for growth. How much water vapor is that? From the ideal gas law we find that for $V = 1$ m^3

$$m_\text{v} = \frac{e_\text{sw} V}{R_\text{v} T} = 0.0175 \text{ kg}.$$

Assuming that a typical raindrop has a radius of 0.5 mm it follows that the mass of a raindrop is $m_\text{raindrop} = V/\rho \approx$ 0.00052 grams. Given the total amount of water vapor available this translates to $\sim 33\,650$ raindrops. However, in 1 m^3 of air there are 10^8 "competitors" with equal rights. Simply, none of them will grow to be a raindrop by condensation only. Other processes must also take place (collision, coalescence, for example).

(6.3) Consider a sample of 2 moles of supercooled water (i.e. liquid water existing at environmental temperatures less than 0 °C). Assume that the temperature of the surroundings is -10 °C and that the supercooled water freezes. In this process the latent heat of fusion (freezing) is lost to the surroundings and finally ice and surroundings return to the original temperature. Calculate $\Delta U, \Delta H$, and ΔS for both supercooled water and surroundings.

Two moles of water equal 36 grams. If we neglect volume and pressure variations during phase changes between water and ice it follows from the first law and the definition of enthalpy that for both supercooled water and surroundings

$$\Delta H = \Delta U = Q.$$

There is a correct and a wrong way to calculate Q and ΔS during this process. If we calculate the amount of heat given away from the supercooled water as it freezes at $-10\,°$C, we find that $Q_\text{w} = -ml_\text{f}(-10\,°\text{C}) = 0.036 \times 0.312 \times 10^6$ J $= -11\,232$ J. Then, if we use equation (5.7) we will find that $\Delta S_\text{w} = \int \delta Q/T = -42.7$ J K^{-1}. The amount of heat released is gained by the surroundings. Thus, $Q_\text{S} = 11\,232$ and $\Delta S_\text{S} = 42.7$. It follows that the change in entropy of the whole system (surroundings + supercooled water) is zero. But *there was* a change in the system, which according to the second law should increase the entropy of the system. The reason for this violation is that we have indirectly assumed that the spontaneous freezing is a reversible process, which is not.

A correct way would be the following: Start with the supercooled water at $-10\,°$C and warm it reversibly to $0\,°$C (step 1), freeze it reversibly at $0\,°$C (step 2) and then cool it reversibly to $-10\,°$C (step 3). In this scenario we have

Step 1

$$Q_1 = mc_{pw}\Delta T = 1518.5 \text{ J}$$

$$\Delta S_1 = \int_{T_1}^{T_2} mc_{pw}\frac{dT}{T} = 5.67 \text{ J K}^{-1}$$

Step 2

$$Q_2 = -ml_f(0 \text{ °C}) = -12\,013.2 \text{ J}$$

$$\Delta S_2 = \frac{Q_2}{T_2} = -44.0 \text{ J K}^{-1}$$

Step 3

$$Q_3 = mc_{pi}\Delta T = -758.2 \text{ J}$$

$$\Delta S_3 = \int_{T_2}^{T_1} mc_{pi}\frac{dT}{T} = -2.83 \text{ J K}^{-1}$$

In the above calculations we have considered that $T_1 = 263$ K, $T_2 = 273$ K, $c_{pw} = c_w = 4218$ J kg^{-1} K^{-1}, $c_{pi} = c_i = 2106$ J kg^{-1} K^{-1}, and $l_f(0 \text{ °C}) = 0.3337 \times 10^6$ J kg^{-1} K^{-1}. It follows that the amount of heat released by the supercooled water is

$$Q_w = Q_1 + Q_2 + Q_3 = -11\,252.9 \text{ J},$$

and the change in entropy of the supercooled water is

$$\Delta S_w = \Delta S_1 + \Delta S_2 + \Delta S_3 = -41.16 \text{ J K}^{-1}.$$

The surroundings ($T_S = -10 \text{ °C}$) gain the amount that the freezing of supercooled water releases. Thus,

$$\Delta U_S = \Delta H_S = Q_S = -Q_w = 11\,252.9 \text{ J}.$$

At the same time their change of entropy is

$$\Delta S_S = \int \frac{\delta Q_S}{T_S} = \frac{Q_S}{T_S} = 42.79 \text{ J K}^{-1}.$$

It follows that the change in the entropy of the whole system (surroundings + supercooled water) is

$$\Delta S = \Delta S_S + \Delta S_w = 1.63 \text{ J K}^{-1} > 0.$$

Problems

(6.1) Prove that for any system

$$\left(\frac{\partial U}{\partial V}\right)_T = T\left(\frac{\partial p}{\partial T}\right)_V - p.$$

(Hint: express dU using equation (2.1) and substitute into the first law. Then consider the definition of $dS = \delta Q/T$

and the fact that dS is an exact differential, perform the partial differentiations and collect terms.)

(6.2)   Using the equation derived in problem 6.1, show that for an ideal gas U is only a function of temperature and does not depend on V (Joule's law).

(6.3)   By plotting e_{sw} against T show that in the range $-20\,°C$ to $30\,°C$ equations (6.16) and (6.17) are virtually identical. If we were very picky about choosing one over the other which would you choose?

(6.4)   Inside an open container is 500 g of ice with temperature $T = -20\,°C$. Heat is then provided to the ice at a rate of 100 cal min^{-1} and for 700 minutes. Plot (a) the relationship between temperature (in $°C$) and time (in seconds), (b) the relationship between the heat absorbed by the ice and time, (c) the relationship between the heat absorbed and temperature. Neglect the heat capacity of the container and assume that $l_f = \text{constant} = 79.7$ cal g^{-1} and $c_i = \text{constant} = 0.503$ cal g^{-1}K^{-1}.

(6.5)   A closed box of volume 2 m^3 and temperature 120 $°C$ contains 1 m^3 of water saturated with vapor. Calculate (a) the mass of water vapor, (b) the mass of water vapor that must slowly escape in order for the temperature inside the box to decrease to 100 $°C$ while maintaining saturated conditions. (1.43 kg, 0.71 kg)

(6.6)   An open container of volume 2 liters contains dry air and some quantity of water at standard conditions. After equilibrium is achieved the container is closed and heated to a temperature of 100 $°C$. What is the least amount of water that must have been added if after the heating the air is saturated? What is the final pressure inside the box? (1.43 g, 2.6 atmospheres)

(6.7)   A substance obeys the equation of state $pV^{1.2} = 10^9 T^{1.1}$. A measurement of its internal capacity inside a container having a constant volume of 100 liters shows that under these conditions the thermal capacity is constant and equal to 0.1 cal K^{-1}. Express the energy and the entropy of the system as function of T and V. (Hint: is the substance an ideal gas?)

(6.8)   How would you have to alter your assumptions in example 6.2 in order to make cloud droplets grow to raindrops by condensation only?

(6.9)   Calculate the change in entropy when 5 kg of water at 10 $°C$ is raised to 100 $°C$ and then converted into steam at that temperature. At 100 $°C$ the latent heat of vaporization of water is 2.253×10^6 J kg^{-1}. In the above range of temperatures, the specific heat capacity under constant volume for

Figure 6.4
Illustration for problem
6.10.

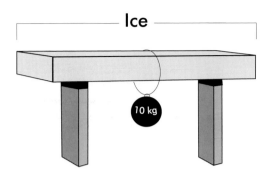

water is nearly independent of temperature with a value of about 4.18×10^3 J kg^{-1} K^{-1}. ($\geq 35\,972$ J kg^{-1})

(6.10) A block of ice rests as shown in figure 6.4. A wire on which we attach a weight surrounds the ice. Explain why the wire will go through the ice block without the ice breaking into two pieces.

(6.11) The warm source of a Carnot cycle has a temperature of 100 °C while the cold source consists of melting ice. When the cycle operates for one hour it is observed that 1 ton of ice has melted. Calculate (1) the heat absorbed in the cycle from the warm source (2) the heat given away in the cycle to the cold source. The latent heat of fusion at 100 °C is 0.3337×10^6 J kg^{-1}. (333.7×10^6 J, 244.2×10^6 J)

CHAPTER SEVEN

Moist air

Our atmosphere is basically a two-component system. One component is dry air and the other is water existing in vapor and possibly one of the condensed phases (liquid water or ice). According to Dalton's law, in a mixture of ideal gases each gas can be assumed to behave as if the other gases were absent. As such, in a mixture of dry air, water vapor, and a condensed phase, the "water" system (water vapor + condensed phase) can be treated as being independent of the dry air. In this case, the concepts developed in the previous chapter (for the one-component heterogeneous system "water") are valid for the two-component heterogeneous system "dry air + water". We will call the system consisting of dry air and water vapor "moist air" and it can be unsaturated or saturated with water vapor. Since liquid water is absent, moist air is a two-component system with one phase present. As such, according to equation (6.1) we need three state variables to specify the system's state. Usually these variables are taken to be pressure, temperature, and a new variable (to be defined soon) called mixing ratio. If the condensed phase is present and in equilibrium with the vapor phase, then two variables are needed (typically temperature and pressure). For clarity in our notation we will use the subscripts d, w, v to indicate dry air, liquid water and water vapor respectively. The only exception will be for the vapor pressure, which we will denote simply as e. Variables with no subscripts will correspond to a mixture of dry air and water.

7.1 Measures and description of moist air

7.1.1 Humidity variables

In a sample of moist air, dry air and water vapor have the same temperature T and occupy the same volume V. Thus, for water vapor

$$eV = m_{\mathrm{v}} R_{\mathrm{v}} T$$

or

$$e = \rho_{\mathrm{v}} R_{\mathrm{v}} T$$

where e is the vapor pressure, ρ_{v} is the density of water vapor, and R_{v} is the specific gas constant for water vapor. Since the molecular weight of water $M_{\mathrm{v}} = 18.01$ g mol^{-1}, the value of R_{v} is equal to $R^{*}/M_{\mathrm{v}} = 461.5$ J kg^{-1} K^{-1}. Since $R^{*} = R_{\mathrm{d}} M_{\mathrm{d}}$ it follows that $R_{\mathrm{d}} M_{\mathrm{d}} = R_{\mathrm{v}} M_{\mathrm{v}}$. Thus,

$$\epsilon = \frac{R_{\mathrm{d}}}{R_{\mathrm{v}}} = \frac{M_{\mathrm{v}}}{M_{\mathrm{d}}} = 0.622. \tag{7.1}$$

We define the specific humidity, q, and mixing ratio, w, as

$$\mathsf{q} = \frac{\rho_{\mathrm{v}}}{\rho} = \frac{m_{\mathrm{v}}}{m}$$

and

$$w = \frac{\rho_{\mathrm{v}}}{\rho_{\mathrm{d}}} = \frac{m_{\mathrm{v}}}{m_{\mathrm{d}}} \tag{7.2}$$

where $m = m_{\mathrm{d}} + m_{\mathrm{v}}$ is the total mass of the mixture (ρ_{v} is also called the absolute humidity). Since $\rho_{\mathrm{v}} = eM_{\mathrm{v}}/R^{*}T$, $\rho_{\mathrm{d}} = p_{\mathrm{d}} M_{\mathrm{d}}/R^{*}T$, and $p_{\mathrm{d}} = p - e$ it follows that

$$w = \epsilon \frac{e}{p - e}. \tag{7.3}$$

At saturation the *saturation* mixing ratio is

$$w_{\mathrm{s}} = \epsilon \frac{e_{\mathrm{s}}}{p - e_{\mathrm{s}}} \tag{7.4}$$

where e_{s} is either the equilibrium (saturation) vapor pressure with respect to liquid water (e_{sw}) or with respect to ice (e_{si}). In general $e_{\mathrm{s}}, e \ll p$ which reduces the above two equations to

$$w \approx \epsilon \frac{e}{p}, \quad w_{\mathrm{s}} \approx \epsilon \frac{e_{\mathrm{s}}}{p}. \tag{7.5}$$

Since $1/\mathsf{q} = (m_{\mathrm{d}} + m_{\mathrm{v}})/m_{\mathrm{v}} = (m_d/m_v) + 1$ it follows that

$$\frac{1}{\mathsf{q}} = \frac{1}{w} + 1$$

or

$$w = \frac{\mathsf{q}}{1 - \mathsf{q}}, \quad \mathsf{q} = \frac{w}{1 + w}. \tag{7.6}$$

In the atmosphere both w and q are very small ($w, \mathsf{q} \ll 1$). For this reason we can always assume that $w \approx \mathsf{q}$.

The relative humidity, r, is defined as

$$r = \frac{m_{\mathrm{v}}}{m_{\mathrm{vs}}}$$

where m_v is the mass of water vapor in a sample of moist air of volume V and m_{vs} is the mass of water vapor the sample would have had if it were saturated. Because of the ideal gas law we can write r as

$$r = \frac{e}{e_s}. \tag{7.7}$$

Using (7.5) it follows that approximately

$$r \approx \frac{w}{w_s}.$$

7.1.2 Mean molecular weight of moist air and other quantities

According to equation (3.13) we have that the mean molecular weight of moist air is

$$\overline{M} = \frac{m_d + m_v}{\frac{m_d}{M_d} + \frac{m_v}{M_v}}.$$

We can manipulate the above equation by writing it as

$$\frac{1}{\overline{M}} = \left(\frac{m_d}{M_d} + \frac{m_v}{M_v} \right) \left(\frac{1}{m_d + m_v} \right)$$

or

$$\frac{1}{\overline{M}} = \frac{1}{M_d} \left[\frac{1}{m_d + m_v} \left(m_d + \frac{m_v M_d}{M_v} \right) \right]$$

or

$$\frac{1}{\overline{M}} = \frac{1}{M_d} \frac{m_d}{m_d + m_v} \left(1 + \frac{m_v/m_d}{M_v/M_d} \right)$$

or

$$\frac{1}{\overline{M}} = \frac{1}{M_d} \frac{1}{1 + w} \left(1 + \frac{w}{\epsilon} \right)$$

or

$$\frac{1}{\overline{M}} = \frac{1}{M_d} \left(\frac{1}{1 + w} + \frac{q}{\epsilon} \right)$$

or

$$\frac{1}{\overline{M}} = \frac{1}{M_d} \left(1 - q + \frac{q}{\epsilon} \right)$$

or

$$\frac{1}{\overline{M}} = \frac{1}{M_d} \left[1 + \left(\frac{1}{\epsilon} - 1 \right) q \right]$$

or

$$\frac{1}{\overline{M}} = \frac{1}{M_d} (1 + 0.61q).$$

Then for moist air the equation of state is

$$pa = R_{\text{moist}}T = \frac{R^*}{M}T = \frac{R^*}{M_{\text{d}}}(1 + 0.61q)T,$$
$$= R_{\text{d}}(1 + 0.61q)T. \tag{7.8}$$

The above equation defines the *virtual temperature*,

$$T_{\text{virt}} = (1 + 0.61q)T, \tag{7.9}$$

which can be interpreted as the temperature of dry air having the same values of p and a as the moist air. Otherwise stated, virtual temperature is the temperature that air of a given pressure and volume (or density) would have if the air were completely free of water vapor. Since q in reality is always greater than zero, T_{virt} is always greater than T. The above equation also gives the gas constant of the mixture:

$$R = (1 + 0.61q)R_{\text{d}}. \tag{7.10}$$

Similarly, we can define the specific heat capacities of moist air. Suppose that a sample of such air receives, at a constant pressure, an amount of heat δQ which increases its temperature by dT. Some of this amount (δQ_{d}) is absorbed by the dry air and some (δQ_{v}) by the water vapor. In this case we can write that

$$\delta Q = \delta Q_{\text{d}} + \delta Q_{\text{v}}$$

or

$$\delta Q = m_{\text{d}}\delta q_{\text{d}} + m_{\text{v}}\delta q_{\text{v}}$$

or

$$\delta q = \frac{m_{\text{d}}}{m_{\text{d}} + m_{\text{v}}}\delta q_{\text{d}} + \frac{m_{\text{v}}}{m_{\text{d}} + m_{\text{v}}}\delta q_{\text{v}}$$

or

$$\delta q = (1 - q)\delta q_{\text{d}} + q\delta q_{\text{v}}.$$

Recalling the definition of specific heat capacities we then have

$$c_p = (1 - q)\frac{\delta q_{\text{d}}}{dT} + q\frac{\delta q_{\text{v}}}{dT}$$

or

$$c_p = (1 - q)c_{pd} + q\, c_{pv}$$

or

$$c_p = c_{pd}\left[1 + \left(\frac{c_{pv}}{c_{pd}} - 1\right)q\right]$$

or

$$c_p = c_{pd}(1 + 0.87q) \approx c_{pd}(1 + 0.87w). \tag{7.11}$$

Similarly

$$c_V = (1 - q)c_{Vd} + qc_{Vv}$$

which leads to

$$c_V = c_{Vd}(1 + 0.97q) \approx c_{Vd}(1 + 0.97w). \qquad (7.12)$$

From equations (7.11) and (7.12) we also have

$$\gamma = \frac{c_p}{c_V} = \frac{c_{pd}}{c_{Vd}} \frac{(1 + 0.87q)}{(1 + 0.97q)},$$

which because $q \ll 1$ can be approximated with

$$\gamma \approx \gamma_d(1 + 0.87q)(1 - 0.97q)$$
$$\approx \gamma_d(1 - 0.1q), \qquad (7.13)$$

where $\gamma_d = 1.4$. Similarly,

$$k = \frac{\gamma - 1}{\gamma} = \frac{R}{c_p} = \frac{R_d}{c_{pd}} \frac{(1 + 0.61q)}{(1 + 0.87q)}$$
$$\approx k_d(1 + 0.61q)(1 - 0.87q)$$
$$\approx k_d(1 - 0.26q), \qquad (7.14)$$

where $k_d = 0.286$. Note that similar expressions in the presence of liquid water cannot be derived because liquid water + vapor does not constitute an ideal gas.

7.2 Processes in the atmosphere

In this section we will discuss scenarios that are relevant to basic processes in the atmosphere. To make the discussion flow effectively we will start with processes happening at a constant pressure level (for example, at the surface). After that we will proceed with processes involving ascent, first of unsaturated air and then of saturated air. Finally, we will discuss processes that involve mixing of two air masses in the horizontal and in the vertical. In the following presentation when a condensed phase is present, we will assume that it is liquid water. The same theory holds when the condensed phase is ice as long as the appropriate substitutions are made (for example e_{si} for e_{sw}, l_s for l_v, c_i for c_w, etc.).

7.2.1 Isobaric cooling – dew and frost temperatures

Consider a parcel of *unsaturated* moist air and assume it is a closed system. In this case q and w remain constant. If this parcel begins to cool at a constant pressure the vapor pressure e will also remain constant as long as condensation does not occur (recall that $e = wp/(w + \epsilon)$). However, the equilibrium vapor pressure will not remain constant. As we saw in the last chapter both e_{sw} and e_{si} are strongly dependent on T, and as T decreases so does e_s. Since e_s

Figure 7.1
Graphical definition of the
dew point temperature,
T_{dew}. Clearly,
$e_{\mathrm{sw}}(T_{\mathrm{dew}}) = e(T)$.

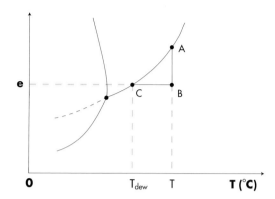

decreases it follows that the relative humidity increases and saturation is approached. The temperature at which saturation is reached
is called the dew point temperature, T_{dew}, if saturation is reached
with respect to liquid water or the frost point temperature, T_{f}, if
saturation is reached with respect to ice.

These temperatures derive their name from the fact that if the
parcel cools isobarically *below* these temperatures droplets (or ice)
will form. However, keep in mind that for condensation (or sublimation) to happen at temperatures just below saturation point either
condensation nuclei or solid surfaces must be present. Spontaneous
condensation in the absence of nuclei or surfaces needs extremely
high relative humidities – supersaturation – which cannot easily
occur in the atmosphere.

Isobaric cooling can happen naturally because of radiation cooling of the surface or the air itself. In this case the ground can lose
enough heat to cool below the temperature at which atmospheric
water vapor condenses. If this happens at temperatures greater
than $0\,^{\circ}\mathrm{C}$ water vapor in the layer very close to the surface condenses on the surface and becomes visible as dew. If it happens at
temperatures less than $0\,^{\circ}\mathrm{C}$ the product is frost. If the cooling is
very strong or if the air itself cools by radiation, the layers of air
above the surface may also cool below their saturation point lead-
ing to the formation of radiation fog. Alternatively, isobaric cooling
can be achieved when air moves in the horizontal ($p \approx$ constant)
and travels over colder regions. Then saturation may occur as the
air cools by heat conduction to the surface. This leads to advection
fogs.

At T_{dew} vapor pressure becomes the equilibrium (saturation)
vapor pressure (and to a very good approximation the mixing ratio
becomes the saturation mixing ratio). As such if e is the partial
vapor pressure at T, the dew point must satisfy the equation

$$e_{\mathrm{sw}}(T_{\mathrm{dew}}) = e. \qquad (7.15)$$

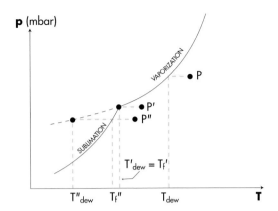

Figure 7.2
Depending on where a given state is in a (p, T) diagram, T_{dew} or T_f may or may not be defined. For example, for state P only T_{dew} can be defined.

This is shown graphically in figure 7.1 where point B is assumed to have a temperature T and vapor pressure e. To approach saturation isobarically we must draw a line from B perpendicular to the pressure axis. Point C corresponds to a temperature where the water vapor pressure is its equilibrium vapor pressure. This point thus defines T_{dew}. In this diagram $AT = e_{\mathrm{sw}}(T)$ and $BT = CT_{\mathrm{dew}} = e = e_{\mathrm{sw}}(T_{\mathrm{dew}})$. Obviously, since the equilibrium vapor pressure curve is known the relative humidity and dew point temperature of any point in the diagram can easily be found ($r = TB/TA$, $T_{\mathrm{dew}} = OT - BC$). If only the temperature is known then we need to know T_{dew} in order to estimate the relative humidity and the absolute humidity. From the figure it follows that neither relative humidity nor the *dew point spread* (or dew point depression), $T - T_{\mathrm{dew}}$, are direct measures of the actual water vapor amount. Both describe how far we are from the equilibrium curve but similar values for r or $T - T_{\mathrm{dew}}$ correspond to different e and thus to different ρ_v (absolute humidity). Note that, depending on where a given point in the (e, T) diagram is, T_{dew} (or T_f) may or may not be defined. For example, in figure 7.2, given point P only T_{dew} can be defined. For point P', $T'_{\mathrm{dew}} = T'_f = $ triple point temperature. For point P'' and in the presence of supercooled water, both T''_f and T''_{dew} can be defined with $T''_f > T''_{\mathrm{dew}}$. If supercooled water is not present, T''_{dew} cannot be defined.

Estimating T_{dew}

Since $e = e_{\mathrm{sw}}(T_{\mathrm{dew}})$ the C–C equation can be written as

$$\frac{de}{dT_{\mathrm{dew}}} = \frac{l_v e}{R_v T_{\mathrm{dew}}^2}$$

or

$$\frac{de_{\mathrm{sw}}}{e_{\mathrm{sw}}} = \frac{l_v dT}{R_v T^2}.$$

Considering $l_v \approx$ constant and integrating from T_{dew} (where $e_{sw}(T_{dew}) = e$) to T (where $e_{sw}(T) = e_{sw}$) we obtain

$$\ln \frac{e_{sw}}{e} = \frac{l_v}{R_v}\left(\frac{T - T_{dew}}{TT_{dew}}\right)$$

or

$$-\ln r = \frac{l_v}{R_v}\left(\frac{T - T_{dew}}{TT_{dew}}\right)$$

or

$$T - T_{dew} = -R_v TT_{dew} \, \ln r/l_v. \qquad (7.16)$$

Equation (7.16) provides T_{dew} from temperature and relative humidity. Using $l_v \approx 2.501 \times 10^6$ J kg^{-1}, $R_v = 461.51$ J kg^{-1}K^{-1}, the above equation becomes

$$T - T_{dew} = -1.845 \times 10^{-4} TT_{dew} \, \ln r.$$

Following the same procedure we can obtain (for $l_s = 2.8345$ J kg^{-1})

$$T - T_f = -1.628 \times 10^{-4} TT_f \, \ln r.$$

7.2.2 Adiabatic isobaric processes – wet-bulb temperatures

Let us now suppose that our system is unsaturated moist air and liquid water and that the mixture has a temperature T and a combined water mixing ratio $w_t = (m_v + m_w)/m_d$. Since this is an unstable situation the system will tend to approach equilibrium. This will be achieved by water evaporating. If we assume that the system is closed and the process is taking place adiabatically (i.e. the process is also isenthalpic), then the liquid water must draw heat for the evaporation from the unsaturated moist air. Since no other processes are taking place this decrease in temperature must be related to the change in the mixing ratio of moist air due to evaporating liquid water. Note that, since at any time during this process the system differs slightly from saturation, the process is spontaneous and irreversible. If there is enough liquid water, saturation will be reached. The temperature at saturation is called the *isobaric wet-bulb temperature*, T_w. An important atmospheric process described by this type of transformation is cooling by evaporation of rain at a given level. The system consists of a certain air mass plus water that evaporates into it from the rain falling through it. If instead of water we had ice present, then we would define the isobaric ice-bulb temperature, T_i. Note that T_w is not the same as T_{dew} since the two cooling processes are different. Now, if we follow the reverse procedure, we could define the temperature which air would achieve if all its vapor were to condense in an adiabatic isobaric process and with the produced water being withdrawn. This limiting temperature is called the isobaric equivalent temperature, T_{ei}. Note, however, that the definition of this temperature is *not* a "legal" one. The reason is that we assume that the reverse of an irreversible process can happen, which is not possible.

Derivation of the wet-bulb temperature

Consider a closed heterogeneous system consisting of dry air, water vapor and liquid water. If this system undergoes a change, then the change in enthalpy is

$$\Delta H = H_f - H_i$$

where the subscripts i and f refer to the initial and final state, respectively. We can write the above equation in terms of the three components

$$\Delta H = (H_{df} + H_{vf} + H_{wf}) - (H_{di} + H_{vi} + H_{wi})$$

or

$$\Delta H = (m_d h_{df} + m_{vf} h_{vf} + m_{wf} h_{wf}) - (m_d h_{di} + m_{vi} h_{vi} + m_{wi} h_{wi}). \tag{7.17}$$

If we recall the definition of l_v ($l_v = h_v - h_w$) and consider that the total amount of water $m_t = m_v + m_w$ is conserved (i.e. $m_{vi} + m_{wi} = m_{vf} + m_{wf} = m_t$), then the above equation becomes

$$\Delta H = m_d(h_{df} - h_{di}) + m_t(h_{wf} - h_{wi}) + m_{vf} l_{vf} - m_{vi} l_{vi}. \tag{7.18}$$

If we now assume that in the range of meteorological interest c_{pd} and c_w are nearly independent of temperature, then we can write (recall equation (4.18)) that

$$h_{df} - h_{di} = c_{pd}(T_f - T_i)$$
$$h_{wf} - h_{wi} = c_w(T_f - T_i).$$

In addition, if T_i is not greatly different from T_f we can assume that

$$l_{vi} = l_{vf} = l_v.$$

Then equation (7.18) becomes

$$\Delta H = m_d c_{pd}(T_f - T_i) + m_t c_w(T_f - T_i) + (m_{vf} - m_{vi}) l_v$$

or

$$\Delta H / m_d = \left(c_{pd} + \frac{m_t}{m_d} c_w \right)(T_f - T_i) + \left(\frac{m_{vf} - m_{vi}}{m_d} \right) l_v$$

or

$$\Delta H / m_d = (c_{pd} + w_t c_w)(T_f - T_i) + (w_f - w_i) l_v$$

or

$$\Delta H / m_{\rm d} = (c_{p{\rm d}} + w_{\rm t} c_{\rm w})\Delta T + l_{\rm v}\Delta w. \qquad (7.19)$$

Note that $w_{\rm t}$ (the total water mixing ratio) is a different constant for each system and varies only with the concentration of liquid water. As such it does not depend on temperature. Since we have also assumed that $c_{p{\rm d}}, c_{\rm w}, l_{\rm v} \neq f(T)$ we can write the above equation in differential form:

$$dH / m_{\rm d} = (c_{p{\rm d}} + w_{\rm t} c_{\rm w})dT + l_{\rm v}dw. \qquad (7.20)$$

Equation (7.20) is a general expression for the change in enthalpy of a closed heterogeneous system of dry air, water vapor and liquid water during any process. If we restrict ourselves to isenthalpic processes only, then equation (7.20) becomes

$$(c_{p{\rm d}} + w_{\rm t} c_{\rm w})(T_i - T_f) = (w_f - w_i)l_{\rm v}. \qquad (7.21)$$

Equation (7.21) indicates that in an adiabatic isobaric process, where moist air cools from evaporating liquid water, the change in temperature relates to the change in the mixing ratio. It also tells us that T_f depends on $w_{\rm t}$ with T_f becoming minimum when $w_{\rm t} \to 0$. Since $m_{\rm t} = m_{\rm v} + m_{\rm w}$ cannot be zero (if it were neither the mixing ratio nor temperature would change), this condition requires that $m_{\rm d} \to \infty$. Of course this cannot be true but to a good approximation $m_{\rm d} \gg m_{\rm t}$. We define this limiting temperature as the wet-bulb temperature, $T_{\rm w}$, which is determined by setting in equation (7.21) $T_i = T, T_f = T_{\rm w}, w_i = w, w_f = w_{\rm sw}$, and $w_{\rm t} = 0$:

$$c_{p{\rm d}}(T - T_{\rm w}) = (w_{\rm sw} - w)l_{\rm v}$$

or

$$T_{\rm w} + \frac{l_{\rm v}}{c_{p{\rm d}}}w_{\rm sw} = T + \frac{l_{\rm v}}{c_{p{\rm d}}}w \qquad (7.22)$$

where $w_{\rm sw}$ is the saturation mixing ratio over liquid water at *temperature* $T_{\rm w}$. As expected, since $w_{\rm sw} > w, T_{\rm w} < T$.

Estimation of the isobaric equivalent temperature

In order to estimate this temperature ($T_{\rm ei}$), the only thing we have to do is to set in equation (7.21) $w_i = w_{\rm t} = w$ (the mixing ratio of the moist air), $w_f = 0, T_i = T$ and $T_f = T_{\rm ei}$. Then we obtain

$$T_{\rm ei} = T + \frac{l_{\rm v}w}{c_{p{\rm d}} + wc_{\rm w}}. \qquad (7.23)$$

Obviously $T_{\rm ei} > T$. Also, since for typical values of w (around 10 g kg^{-1}) $l_{\rm v}/(c_{p{\rm d}} + wc_{\rm w}) \approx 2500$ it follows that $T_{\rm ei} > T_{\rm virt}$.

Relation between wet-bulb and dew point temperatures

If we employ the general approximation $w \approx \epsilon e/p$ we can write equation (7.21) as

$$c_p(T_i - T_f) = \frac{\epsilon}{p}(e_f - e_i)l_v$$

where $c_p = c_{pd} + w_t c_w$. Setting $T_i = T, T_f = T_w$ and $e_f = e_{sw}(T_w)$ yields

$$e_i = e_{sw}(T_w) - \frac{pc_p}{l_v\epsilon}(T - T_w).$$

From the definition of dew point temperature we have that $e_i = e_{sw}(T_{dew})$. Then, the above equation becomes

$$e_{sw}(T_{dew}) = e_{sw}(T_w) - \frac{pc_p}{l_v\epsilon}(T - T_w).$$

Since $T - T_w > 0$ it follows that $e_{sw}(T_{dew}) < e_{sw}(T_w)$. Then, from equation (6.17) it follows that

$$T_{dew} < T_w.$$

Are you keeping up with all these temperatures? Up to now we have discussed five different temperatures: T, temperature; T_{virt}, virtual temperature; T_{dew}, dew point temperature; T_w, isobaric wet-bulb temperature; and T_{ei}, isobaric equivalent temperature. They are related as follows:

$$T_{dew} < T_w < T < T_{virt} < T_{ei}. \tag{7.24}$$

Brace yourself. There are more!

7.2.3 Adiabatic expansion (or compression) of unsaturated moist air

Recall Poisson's equation for an adiabatic expansion or compression of an ideal gas from p, T to p', T'

$$T' = T\left(\frac{p'}{p}\right)^k$$

which for $p' = 1000$ mbar defined the potential temperature, θ,

$$\theta = T\left(\frac{1000}{p}\right)^k. \tag{7.25}$$

Poisson's equation is valid for any ideal gas. As such it is valid for moist air provided that the appropriate k is used and that condensation does not take place. We can then define the potential temperature of unsaturated moist air as

$$\theta_m = T\left(\frac{1000}{p}\right)^{k_d(1-0.26q)}. \tag{7.26}$$

If we substitute T from equation (7.25) applied to dry air into equation (7.26) we obtain (assuming that $p_d \approx p$)

$$\theta_m = \theta_d \left(\frac{1000}{p} \right)^{-k_d 0.26q}$$

or

$$\theta_m = \theta_d \left(\frac{1000}{p} \right)^{-0.07q}. \tag{7.27}$$

As we have already discussed, in our atmosphere $q \ll 1$ which indicates that to a very good approximation

$$\theta_m \approx \theta_d.$$

As in the case of virtual temperature, here again we can define the *virtual potential temperature*, θ_{virt}, by substituting T_{virt} for T in equation (7.25):

$$\theta_{virt} = T_{virt} \left(\frac{1000}{p} \right)^{k_d}. \tag{7.28}$$

This will be the temperature dry air will acquire if it were to expand or compress from a level (T_{virt}, p) to the 1000 mbar level. Since $T_{virt} > T$ it follows that $\theta_{virt} > \theta$.

Note that even though q and w are very small (of the order of 0.01) we should not jump to the conclusion that all variables or parameters of unsaturated moist air are for all practical purposes the same as those of dry air. If we consider a typical value of $q = w = 0.01$ we find that differences in virtual temperature of the order of 2–3 °C are not uncommon. Such differences are often not negligible. On the other hand, as we saw above, differences in moist potential temperature are indeed negligible (of the order of 0.1 °C). Consequently, from now on the symbol θ will be used to indicate the potential temperature of both dry and unsaturated moist air.

Similarly to the dry adiabatic lapse rate we may define here the moist (unsaturated) adiabatic lapse rate as

$$\Gamma_m = \frac{g}{c_p} = \frac{g}{c_{pd}(1 + 0.87w)}$$

or

$$\Gamma_m = \frac{\Gamma_d}{1 + 0.87w} \approx \Gamma_d(1 - 0.87w). \tag{7.29}$$

7.2.4 Reaching saturation by adiabatic ascent

As an unsaturated moist air parcel rises adiabatically it cools. During the ascent (and as long as no condensation occurs) its q or w remains the same but the equilibrium vapor pressure decreases. At the same time the vapor pressure e also decreases. This can be easily seen by recalling that $e = wp/(w + \epsilon)$. As w and ϵ remain constant e/p remains constant. Since during the ascent p decreases, e must decrease if e/p must remain constant. Accordingly, when unsaturated moist air ascends adiabatically, its relative humidity increases only if e decreases at a slower rate than e_s. If this condition is not met clouds will not form in ascent but they could still form in descent. However, in our experience, clouds never do form as a result of descending air, so this condition is apparently satisfied. But what is the mathematics and physics behind it?

If we log-differentiate equation (7.7) we get

$$d\ln r = d\ln e - d\ln e_{sw}. \tag{7.30}$$

For moist air, Poisson's equation tells us that $Tp^{\frac{1-\gamma}{\gamma}} = $ constant. Since e/p remains constant, it follows that $Te^{\frac{1-\gamma}{\gamma}} = $ constant. Then, again by log-differentiating, we obtain

$$d\ln T = \frac{\gamma - 1}{\gamma} d\ln e. \tag{7.31}$$

Combining (7.30) and (7.31) with the C–C equation yields

$$d\ln r = \frac{\gamma}{\gamma - 1} d\ln T - \frac{l_v}{R_v T^2} dT. \tag{7.32}$$

The first term on the right-hand side of the above equation accounts for the change in relative humidity due to the decrease in e while the second accounts for the change in relative humidity due to the temperature decrease. As one term is positive and the other negative, the net result can be either positive or negative. This indicates that adiabatic expansion can lead to an increase or to a decrease in r, which in turn implies that the opposite (i.e. adiabatic compression) can also lead to a decrease or an increase in r. This opens the road to bizarre scenarios where clouds can actually form on descent. If we rewrite equation (7.32) as

$$\frac{dr}{dT} = \frac{r}{T} \left(\frac{\gamma}{\gamma - 1} - \frac{l_v}{R_v T} \right),$$

we see that if

$$\frac{\gamma}{\gamma - 1} > \frac{l_v}{R_v T}, \tag{7.33}$$

then $dr < 0$ when $dT < 0$ and $dr > 0$ when $dT > 0$, which would indicate that clouds can form on descent. Assuming the typical values for $\gamma \approx 1.4$, $l_v = 2.5 \times 10^6$ J kg^{-1}, and $R_v = 461.5$ J kg^{-1} K^{-1} we find that inequality (7.33) is satisfied only when $T \geq 1550$ K. Obviously, such a situation does not exist on our planet and clouds can only form on ascent. However, all it takes is an exotic planet whose atmosphere has a condensable gas with smaller l_v and all sorts of things may happen.

Now that we have settled this issue, we can expect that ascending moist unsaturated air will eventually reach a relative humidity of 100%. Immediately after that condensation will take place in order to sustain equilibrium conditions. The level at which saturation of ascending unsaturated moist parcels is achieved is called the *lifting condensation level* (LCL). The temperature at this level, T_{LCL}, is called the *saturation temperature*. By integrating equation (7.32) from an initial state T, T_{dew}, r to the final state $T_{\text{LCL}} = T_{\text{dew,LCL}}$, $r_{\text{LCL}} = 1$ we obtain

$$- \ln r = \frac{\gamma}{\gamma - 1} \ln \frac{T_{\text{LCL}}}{T} + \frac{l_{\text{v}}}{R_{\text{v}}} \left(\frac{1}{T_{\text{LCL}}} - \frac{1}{T} \right). \tag{7.34}$$

Since $r = e/e_{\text{sw}}(T)$ and $e = e_{\text{sw}}(T_{\text{dew}})$ we have from the C–C equation

$$e_{\text{sw}}(T_{\text{dew}}) = 6.11 \exp \left(\frac{l_{\text{v}}}{273 R_{\text{v}}} - \frac{l_{\text{v}}}{R_{\text{v}} T_{\text{dew}}} \right)$$

and

$$e_{\text{sw}}(T) = 6.11 \exp \left(\frac{l_{\text{v}}}{273 R_{\text{v}}} - \frac{l_{\text{v}}}{R_{\text{v}} T} \right).$$

Dividing the above two equations and taking logs yields

$$- \ln r = \frac{l_{\text{v}}}{R_{\text{v}}} \left(\frac{T - T_{\text{dew}}}{T T_{\text{dew}}} \right).$$

Accordingly, equation (7.34) becomes

$$\frac{T - T_{\text{dew}}}{T T_{\text{dew}}} = A \ln \frac{T_{\text{LCL}}}{T} + \left(\frac{1}{T_{\text{LCL}}} - \frac{1}{T} \right) \tag{7.35}$$

where

$$A = \left(\frac{\gamma}{\gamma - 1} \right) \bigg/ \left(\frac{l_{\text{v}}}{R_{\text{v}}} \right).$$

Equation (7.35) can be solved numerically to provide the saturation temperature T_{LCL} from T and T_{dew}. If you would rather not solve this equation you may use the approximation provided by Bolton (1980)

$$T_{\text{LCL}} = \frac{1}{\frac{1}{T - 55} - \frac{\ln r}{2840}} + 55, \tag{7.36}$$

where T is in K, $r = w/w_{\text{sw}}$, $w_{\text{sw}} = \epsilon(e_{\text{sw}}/(p - e_{\text{sw}}))$, $e_{\text{sw}} = 6.11 \exp(19.83 - 5417/T)$, and $T - T_{\text{dew}} = -R_{\text{v}} T T_{\text{dew}} \ln r/l_{\text{v}}$.

A parcel that rises from an initial level where the temperature is T to a level where the temperature is T_{LCL} has expanded. Therefore, the vapor density at LCL is smaller than that at the initial level. From the ideal gas law it then follows that $e(T_{\text{LCL}}) < e(T)$.

Because $T_{\text{LCL}} = T_{\text{dew,LCL}}$ and $e(T) = e_{\text{sw}}(T_{\text{dew}})$ this inequality is equivalent to $e_{\text{sw}}(T_{\text{LCL}}) < e_{\text{sw}}(T_{\text{dew}})$. Then from equation (6.17) it follows that $T_{\text{LCL}} < T_{\text{dew}}$. Thus, relation (7.24) can be extended to include yet one more temperature:

$$T_{\text{LCL}} < T_{\text{dew}} < T_{\text{w}} < T < T_{\text{v}} < T_{\text{ei}}. \tag{7.37}$$

Estimating the height of LCL

The height of LCL for a given parcel (z_{LCL}) depends only on its temperature and relative humidity. At z_{LCL} the parcel's temperature equals its dew point: $T_{\text{LCL}} = T_{\text{dew,LCL}}$. As the parcel is lifted from a reference level z_0 its temperature decreases according to

$$T(z) = T_0 - \Gamma_{\text{d}}(z - z_0) \tag{7.38}$$

where T_0 is the temperature of the parcel of z_0 and we have assumed that $\Gamma_{\text{m}} \approx \Gamma_{\text{d}}$. Similarly, its dew point temperature decreases according to

$$T_{\text{dew}}(z) = T_{\text{dew,0}} - \Gamma_{\text{dew}}(z - z_0). \tag{7.39}$$

Here both Γ_{d} and Γ_{dew} are considered constant. From equation (7.38) and (7.39) we have that

$$T_{\text{LCL}} = T_0 - \Gamma_{\text{d}}(z_{\text{LCL}} - z_0)$$

$$T_{\text{dew,LCL}} = T_{\text{dew,0}} - \Gamma_{\text{dew}}(z_{\text{LCL}} - z_0)$$

or since $T_{\text{LCL}} = T_{\text{dew,LCL}}$,

$$z_{\text{LCL}} - z_0 = \frac{T_0 - T_{\text{dew,0}}}{\Gamma_{\text{d}} - \Gamma_{\text{dew}}}. \tag{7.40}$$

In the above equation we know everything except for Γ_{dew}. In order to estimate Γ_{dew} we proceed as follows. The equation that defines dew point is $e = e_{\text{sw}}(T_{\text{dew}})$. Differentiation with respect to z gives

$$\frac{de_{\text{sw}}(T_{\text{dew}})}{dz} = \frac{de}{dz}$$

or

$$\frac{de_{\text{sw}}(T_{\text{dew}})}{dT_{\text{dew}}} \frac{dT_{\text{dew}}}{dz} = \frac{de}{dz}. \tag{7.41}$$

Now we know that $e = wp/(w + \epsilon)$. If we assume that w remains constant until condensation occurs, we have

$$\frac{de}{dz} = \frac{w}{w + \epsilon} \frac{dp}{dz}$$

or

$$\frac{de}{dz} = \frac{e}{p} \frac{dp}{dz} = \frac{e_{\text{sw}}(T_{\text{dew}})}{p} \frac{dp}{dz}. \tag{7.42}$$

Combining equations (7.41) and (7.42) gives

$$\frac{1}{e_{sw}(T_{dew})}\frac{de_{sw}(T_{dew})}{dT_{dew}}\frac{dT_{dew}}{dz} = \frac{1}{p}\frac{dp}{dz}.$$

After considering the C–C equation we can reduce the above to

$$\frac{l_v}{R_v T_{dew}^2}\frac{dT_{dew}}{dz} = \frac{1}{p}\frac{dp}{dz}. \qquad (7.43)$$

Since the ascent is assumed to be adiabatic, we have from Poisson's equation that

$$d\ln T = \frac{\gamma - 1}{\gamma}d\ln p$$

or

$$\frac{dT}{T dz} = \frac{\gamma - 1}{\gamma}\frac{dp}{pdz}. \qquad (7.44)$$

From equations (7.43) and (7.44) it follows that

$$\frac{l_v}{R_v T_{dew}^2}\frac{dT_{dew}}{dz} = \frac{\gamma}{\gamma - 1}\frac{dT}{T dz}$$

$$= \frac{\gamma}{(\gamma - 1)T}\left(-\frac{g}{c_{pd}}\right)$$

or assuming $\gamma \approx \gamma_d$

$$\frac{l_v}{R_v T_{dew}^2}\frac{dT_{dew}}{dz} = -\frac{g}{R_d T}$$

which then gives

$$\Gamma_{dew} = -\frac{dT_{dew}}{dz} = \frac{g}{\epsilon l_v}\frac{T_{dew}^2}{T}.$$

The value of the right-hand side in the atmosphere ranges from about 1.7 to 1.9 °C km^{-1}. Therefore we can assume that on the average

$$\Gamma_{dew} = 1.8\,°C\ km^{-1}.$$

By dropping the subscript 0 in equation (7.40) it follows that the height of LCL is given by the approximate relationship

$$z_{LCL} - z \approx \frac{T - T_{dew}}{8}. \qquad (7.45)$$

Equations (7.35) or (7.36) and (7.45) give the temperature and height (in km) of the level where clouds begin to form from the temperature and dew point temperature at some reference level z.

7.2.5 Saturated ascent

Once saturation is reached, further ascent results in a further increase in relative humidity, which means that the vapor pressure becomes greater than the equilibrium vapor pressure. At this point, the system will return to equilibrium after the "extra" water vapor condenses and water droplets (or ice depending on the temperature) form on condensation nuclei. From this point on two extreme possibilities may be considered. (1) The condensation products remain in the parcel. This is a reversible process because if we reverse the ascent the products will evaporate. It is also adiabatic as we assume that no heat is exchanged between the parcel and the environment. Since it is reversible and adiabatic it is also isentropic. (2) All condensation products fall out and the parcel consists always of dry air plus saturated water vapor. This makes the parcel an open system. Obviously, this process is neither reversible nor adiabatic and thus it is not isentropic. We will call this process a pseudo-adiabatic process. In reality these two extreme cases may never exist as some condensation products are likely to remain suspended in the parcel. However, because of the conditions imposed the extreme cases are easier to treat. Nevertheless, the study of the extreme cases offers useful insights and approximations for the real cases in between.

Reversible saturated adiabatic process

Since the process is assumed to be reversible, the heterogeneous system of dry air, water vapor, and liquid water (or ice if it is too cold) must be at equilibrium at all times. In this case the total entropy of the system, S, is conserved:

$$S = S_d + S_v + S_w = \text{constant}.$$

If we divide all the terms in the above equation by the mass of dry air we arrive at the following expression

$$S/m_d = s_d + \frac{m_v}{m_d}s_v + \frac{m_w}{m_d}s_w.$$

Recalling that the parcel is saturated ($w = w_{sw}$) and the definition of the total water mixing ratio ($w_t = (m_v + m_w)/m_d$) the above equation can be written as

$$S/m_d = s_d + w_{sw}s_v + (w_t - w_{sw})s_w. \qquad (7.46)$$

For a reversible process

$$dS = \frac{\delta Q}{T} = \frac{dU}{T} + \frac{pdV}{T} = \frac{1}{T}\left(\left(\frac{\partial U}{\partial T}\right)_V dT + \left(\frac{\partial U}{\partial V}\right)_T dV\right) + \frac{p}{T}dV.$$
$$(7.47)$$

In addition,

$$dS = \left(\frac{\partial S}{\partial T}\right)_V dT + \left(\frac{\partial S}{\partial V}\right)_T dV. \tag{7.48}$$

From equations (7.47) and (7.48) it follows that

$$\left(\frac{\partial S}{\partial T}\right)_V = \frac{1}{T}\left(\frac{\partial U}{\partial T}\right)_T \tag{7.49}$$

and

$$\left(\frac{\partial S}{\partial V}\right)_T = \frac{1}{T}\left(\left(\frac{\partial U}{\partial V}\right)_T + p\right). \tag{7.50}$$

For liquid water and water vapor in *equilibrium* at temperature T and pressure p equation (7.50) gives

$$T\frac{S_v - S_w}{V_v - V_w} = \frac{U_v - U_w}{V_v - V_w} + p$$

or

$$T(S_v - S_w) = U_v - U_w + p(V_v - V_w).$$

If we use equation (6.3) the above equation becomes

$$T(S_v - S_w) = H_v - H_w = L_v$$

or

$$T(s_v - s_w) = l_v.$$

Substituting s_v from the above equation into equation (7.46) results in

$$S/m_d = s_d + w_t s_w + \frac{l_v w_{sw}}{T}. \tag{7.51}$$

Now recall that

$$s_d = c_{pd}\ln T - R_d \ln p_d + \text{constant} \quad (p = p_d + e_{sw}(T))$$

and, from equation (5.9),

$$s_w = c_w \ln T + \text{constant}.$$

Accordingly, equation (7.51) becomes

$$S/m_d = (c_{pd} + w_t c_w)\ln T - R_d \ln p_d + \frac{l_v w_{sw}}{T} + \text{constant}. \tag{7.52}$$

Since here S and m_d are conserved, it follows that

$$(c_{pd} + w_t c_w)\ln T - R_d \ln p_d + \frac{l_v w_{sw}}{T} = \text{constant}'. \tag{7.53}$$

As previously, here we can assume that $w_t, c_{pd}, c_w \neq f(T)$. However, w_{sw} through equations (7.5) and (6.17) is a function of T. If we assume that l_v is a function of T (and that may be a good idea here as the parcel may rise to very cold temperatures), then we can write equation (7.53) in differential form as follows:

$$(c_{pd} + w_t c_w)d \ln T - R_d d \ln p_d + d \left(\frac{l_v w_{sw}}{T} \right) = 0. \qquad (7.54)$$

Equations (7.53) and (7.54) describe reversible saturated adiabatic processes. Note again that here T and p correspond to *saturation*.

If we now define θ' as

$$\theta' = T \left(\frac{1000}{p_d} \right)^{R_d/(c_{pd}+w_t c_w)} \qquad (7.55)$$

it follows that

$$(c_{pd} + w_t c_w)d \ln \theta' = (c_{pd} + w_t c_w)d \ln T - R_d d \ln p_d. \qquad (7.56)$$

Combining (7.56) and (7.54) gives

$$(c_{pd} + w_t c_w)d \ln \theta' = -d \left(\frac{l_v w_{sw}}{T} \right)$$

or

$$d \ln \theta' = -d \left[\frac{l_v w_{sw}}{(c_{pd} + w_t c_w)T} \right].$$

Integrating the above equation results in

$$\theta' \exp \left[\frac{l_v w_{sw}}{(c_{pd} + w_t c_w)T} \right] = \text{constant}. \qquad (7.57)$$

The above equation defines the family of curves that describe reversible saturated adiabatic processes. We can evaluate equation (7.57) at a state where all the vapor has condensed ($w_{sw} = 0$). This defines a new temperature, the equivalent potential temperature θ_e given by

$$\theta_e = \theta' \exp \left(\frac{l_v w_{sw}}{(c_{pd} + w_t c_w)T} \right) = \text{constant}. \qquad (7.58)$$

Although θ_e is defined for saturated air it can also be defined for unsaturated air at temperature T, pressure p, and mixing ratio w by considering it to be lifted to the LCL. At LCL we have

$$\theta' = T_{LCL} \left(\frac{1000}{p_d} \right)^{R_d/(c_{pd}+w_t c_w)}$$

(equation (7.55)) and thus

$$\theta_e = T_{LCL} \left(\frac{1000}{p_d} \right)^{R_d/(c_{pd}+w_t c_w)} \exp \left[\frac{l_v(T_{LCL})w}{(c_{pd} + w_t c_w)T_{LCL}} \right]. \qquad (7.59)$$

Note that in the above two expressions $p_d = p_{LCL} - e_{sw}(T_{LCL})$.

Another way to express θ_e for unsaturated air at T, p, w is given by Emanuel (1994):

$$\theta_e = T\left(\frac{1000}{p_d}\right)^{R_d/(c_{pd}+w_t c_w)} r^{-wR_v/(c_{pd}+w_t c_w)}$$
$$\times \exp\left[l_v w/(c_{pd} + w_t c_w)T\right] \qquad (7.60)$$

where r is the relative humidity. Here $p_d = p - e(T)$. For $w = 0$ both (7.59) and (7.60) give $\theta_e = \theta$, the potential temperature.

Recall that the moist potential temperature θ_m is (7.26)

$$\theta_m = T\left(\frac{1000}{p}\right)^{R_d/c_{pd}(1-0.26w)}.$$

Comparison between $\theta_m(\approx \theta)$ and θ' indicates that θ' is only approximately equal to θ. Given the definitions of θ' and θ_e, it follows that the equivalent potential temperature is approximately equal to the potential temperature a parcel would have if it were lifted to very low pressure so that all its water vapor condensed.

Pseudo-adiabatic processes

The description of reversible saturated adiabatic processes depends on the value of w_t. The problem with this dependence is that although w_{sw} can be determined at a given state (p, T), w_t cannot. The liquid water mixing ratio can take on any value and as such at a given (p, T) point there may exist an infinite number of reversible saturated adiabats. If we were to describe such a process in the (p, T) domain this would present a great inconvenience. This problem can be dealt with by defining a new process called a pseudo-adiabatic process. In such a process we do away with liquid water by assuming that it is removed as soon as it is produced. Obviously, our system is now an open system and this process is not reversible. However, we can think of the whole process as a two-step process. During the first step we have a reversible saturated adiabatic process, with condensation of a mass of water (dm_w). During the second step the water produced is leaving the system. For the first step the entropy may be defined as previously but with the liquid water term omitted. Accordingly, we set in equation (7.52) $w_t = w_{sw}$ and we obtain

$$S/m_d = (c_{pd}+w_{sw}c_w)\ln T - R_d\ln p_d + \left(\frac{l_v w_{sw}}{T}\right) + \text{constant.} \quad (7.61)$$

Since in this stage S and m_d are conserved we can write the above equation in differential form

$$d((c_{pd} + w_{sw}c_w)\ln T) - R_d d\ln p_d + d\left(\frac{l_v w_{sw}}{T}\right) = 0. \qquad (7.62)$$

After water forms it is removed immediately, thereby decreasing the entropy of the system. However, this process has no effect on the values of T and p. As such equation (7.62) describes the variation of p and T in a pseudo-adiabatic process. Equation (7.62) is very similar to equation (7.54) but here w_{sw} depends on temperature whereas w_t in equation (7.54) does not. This makes equation (7.62) very difficult to solve analytically. However, numerical techniques can be employed to define, similarly to the case of the equivalent potential temperature, a pseudo-equivalent potential temperature, θ_{ep}. As is shown in Bolton (1980), within $0.3\,°C$, θ_{ep} is given by

$$\theta_{ep} = T \left(\frac{1000}{p}\right)^{0.285(1-0.28w)} \exp\left[w(1+0.81w)\left(\frac{3376}{T_{LCL}} - 2.54\right)\right]$$

(7.63)

where (T, p, w) is any state of the parcel saturated or not (w here is dimensionless and T, T_{LCL} are in K).

The pseudo-equivalent potential temperature can be interpreted as the actual temperature achieved by the parcel under the following thermodynamic processes: (a) pseudo-adiabatic expansion to zero (or to a very low) pressure level at which we can assume that all water vapor has condensed and fallen out, and (b) subsequent dry adiabatic descent to 1000 mbar. Note that according to the definition of θ_e no similar meaning can be attached to θ_e. Nevertheless, the differences in the temperature of a parcel in a reversible saturated adiabatic and in a pseudo-adiabatic process are not great and are often neglected as processes in the atmosphere lie somewhere in between true reversible saturated adiabatic processes and pseudo-adiabatic processes. This can be easily verified if we assume that $w_{sw} \approx w_t \ll c_{pd}$ (i.e. $c_{pd} + w_{sw}c_w \approx c_{pd}$) and $p_d \gg e_{ws}$ (i.e. $p \approx p_d$). In this case equations (7.54) and (7.62) become identical: $c_{pd}d\ln T - R_d d\ln p + d\left(w_{sw}l_v/T\right) = 0$.

7.2.6 A few more temperatures

The dry adiabatic descent from zero pressure level to the 1000 mbar level defined the pseudo-equivalent potential temperature, θ_{ep}. At the original (p, T) level it defines the pseudo-equivalent temperature, T_{ep}. Since the descent is dry adiabatic we have that

$$T_{ep} = \theta_{ep} \left(\frac{p}{1000}\right)^{k_d}.$$

Since $T = \theta\,(p/1000)^{k_d}$ it follows that

$$T_{ep} = T\frac{\theta_{ep}}{\theta}.$$

(7.64)

During the process by which T_{ep} is defined, much more water vapor condenses than it would have if it condensed during an adiabatic isobaric process at an initial level below LCL. Therefore (recall the definition of T_{ei}), if all that vapor were to condense at the initial level it would result in a temperature greater than T_{ei}. Thus, $T_{ep} > T_{ei}$, which makes T_{ep} the highest temperature.

Many more temperatures can be defined but I will spare you this! Just keep in mind that at a given state, if the air is unsaturated, a dry adiabat can be followed to 1000 mbar to define a corresponding potential temperature. As such from the state (T_{ei}, p) the isobaric equivalent potential temperature, θ_{ei}, can be defined. Because $T_{ep} > T_{ei}$ it follows that $\theta_{ep} > \theta_{ei}$.

Similarly, if the air is saturated, a pseudo-adiabat can be followed to either the original pressure level or to the 1000 mbar level to define yet more temperatures. For example, following the pseudo-adiabat corresponding to point (T_{LCL}, p_{LCL}) to (T, p) level defines the pseudo-wet-bulb temperature, T_{wp}, and continuing down to 1000 mbar defines the pseudo-wet-bulb potential temperature θ_{wp}. Obviously, since both θ_{ep} and θ_{wp} are defined by the same curve they must be related and conserved in a pseudo-adiabatic process. θ_{wp} can be estimated from equation (7.63) by setting $T = T_{LCL} = \theta_{wp}$ and $w = w' = w_s(p = 1000 \text{ mbar}, T = \theta_{wp})$, i.e.

$$\theta_{ep} = \theta_{wp} \exp\left[w'(1 + 0.81 w') \left(\frac{3376}{\theta_{wp}} - 2.54 \right) \right]. \qquad (7.65)$$

Similarly, following the pseudo-adiabat passing through point (T_w, p) to 1000 mbar defines the isobaric wet-bulb potential temperature θ_w. Here we should note that T_{wp} and T_w (or θ_{wp} and θ_w) are not exactly the same. T_w and θ_w correspond to a true saturated adiabatic process whereas T_{wp} and θ_{wp} correspond to a pseudo-adiabatic process. Since water remains around in a true saturated adiabatic process it turns out that $T_{wp} < T_w$ and $\theta_{wp} < \theta_w$.

Here is then the final breakdown of all the temperatures and potential temperatures and their relationships:

$$T_{LCL} < T_{dew} < T_{wp} < T_w < T < T_{virt} < T_{ei} < T_{ep}$$

$$(7.66)$$

$$\theta_{wp} < \theta_w < \theta(\approx \theta_m) < \theta_{virt} < \theta_{ei} < \theta_{ep}.$$

7.2.7 Saturated adiabatic lapse rate

Since condensation of water vapor releases heat during a reversible saturated adiabatic ascent, the cooling of the ascending air slows down. As such the corresponding lapse rate should be smaller than that of the dry or moist unsaturated adiabatic ascent.

We could derive an expression for this saturated lapse rate, Γ_s, by starting with equation (7.54)

$$c_p \frac{dT}{T} - R_d \frac{dp_d}{p_d} + d\left(\frac{l_v w_{sw}}{T}\right) = 0$$

where $c_p = c_{pd} + w_t c_w$. Multiplying all terms of the above equation by T/c_p gives

$$dT - \frac{R_d T}{c_p p_d} d(p - e_{sw}) + \frac{1}{c_p} d(l_v w_{sw}) - \frac{l_v w_{sw}}{c_p T} dT = 0,$$

where we have considered that $p = p_d + e_{sw}$. The second term, using the C–C equation in the form $de_{sw}/dT = e_{sw} l_v / R_v T^2$ and the fact that $R_d e_{sw}/R_v p_d = w_{sw}$, can be written as

$$-\frac{R_d T}{c_p p_d} dp + \frac{w_{sw} l_v}{c_p T} dT.$$

Substituting this into the previous equation, dividing all terms by dz, assuming that $dp \approx dp_d$, and using the hydrostatic approximation, leads to

$$\frac{dT}{dz} = -\frac{g}{c_p} - \frac{1}{c_p} \frac{d}{dz}(l_v w_{sw}). \tag{7.67}$$

Note that since in the above equation $c_p \neq c_{pd}$, the term $-g/c_p$ is not exactly the dry adiabatic lapse rate. By performing the differentiation on the right-hand side we can write equation (7.67) as

$$\frac{dT}{dz} = -\frac{g}{c_p} - \frac{w_{sw}}{c_p} \frac{dl_v}{dz} - \frac{l_v}{c_p} \frac{dw_{sw}}{dz}$$

or

$$\frac{dT}{dz} = -\frac{g}{c_p} - \frac{w_{sw}}{c_p} \frac{dl_v}{dT} \frac{dT}{dz} - \frac{l_v}{c_p} \frac{dw_{sw}}{dz}$$

or

$$\left(1 + \frac{w_{sw}}{c_p} \frac{dl_v}{dT}\right) \frac{dT}{dz} = -\frac{g}{c_p} - \frac{l_v}{c_p} \frac{dw_{sw}}{dz}. \tag{7.68}$$

Using equation (6.12) the above equation becomes

$$\left[1 - \frac{w_{sw}(c_w - c_{pv})}{c_p}\right] \frac{dT}{dz} = -\frac{g}{c_p} - \frac{l_v}{c_p} \frac{dw_{sw}}{dz}.$$

Typical values for c_w, c_{pv}, c_p, and w_{sw} are 4220 J kg^{-1} K^{-1}, 1850 J kg^{-1} K^{-1}, 1050 J kg^{-1} K^{-1}, and 0.01 respectively. Such values make the term $w_{sw}(c_w - c_{pv})/c_p$ of the order of 0.02. As such, it can be neglected to arrive at the following approximation

$$\frac{dT}{dz} = -\frac{g}{c_p} - \frac{l_v}{c_p} \frac{dw_{sw}}{dz}. \tag{7.69}$$

Now all we have to do is to find an expression for dw_{sw}/dz. Log-differentiating equation (7.5) leads to

$$\frac{1}{w_{sw}}\frac{dw_{sw}}{dz} = -\frac{1}{p}\frac{dp}{dz} + \frac{1}{e_{sw}}\frac{de_{sw}}{dz}$$

$$= -\frac{1}{p}\frac{dp}{dz} + \frac{1}{e_{sw}}\frac{de_{sw}}{dT}\frac{dT}{dz}$$

where p is the total pressure of the parcel or the surroundings. If we again use the hydrostatic approximation and the C–C equation the above equation becomes

$$\frac{1}{w_{sw}}\frac{dw_{sw}}{dz} = \frac{g}{RT} + \frac{l_v}{R_v T^2}\frac{dT}{dz}. \qquad (7.70)$$

Combining equations (7.70) and (7.69) leads to

$$\Gamma_s = -\frac{dT}{dz} = \frac{g}{c_p}\frac{1 + l_v w_{sw}/RT}{1 + l_v^2 w_{sw}/c_p R_v T^2} \qquad (7.71)$$

where $c_p = c_{pd} + w_t c_w$ and R is the gas constant for the surroundings. In other thermodynamics books you may find some slight differences in the formula giving Γ_s. Most of the differences are due to certain assumptions and approximations. For typical values of saturation mixing ratio and temperature in the lower troposphere, Γ_s ranges around $5\,^{\circ}\mathrm{C}\,\mathrm{km}^{-1}$, which is half the dry adiabatic rate.

7.3 Other processes of interest

7.3.1 Adiabatic isobaric mixing

Let us consider two air masses of temperatures T_1 and T_2 and humidities q_1 and q_2, which are mixing at the same pressure. This corresponds to a horizontal mixing. If we assume the mixing to be an adiabatic process, then it is also isenthalpic. If we assume that no condensation takes place, then we can write for the change in enthalpy of the whole system

$$\Delta H = m_1 \Delta h_1 + m_2 \Delta h_2 = 0 \qquad (7.72)$$

where

$$\Delta h_1 = c_{p1}\Delta T = c_{p1}(T - T_1)$$
$$\Delta h_2 = c_{p2}\Delta T = c_{p2}(T - T_2)$$

and T is the final temperature of the mixture. Considering equation (7.11), equation (7.72) becomes

$$m_1 c_{pd}(1 + 0.87q_1)(T - T_1) + m_2 c_{pd}(1 + 0.87q_2)(T - T_2) = 0.$$

Solving for T we get

$$T = \frac{(m_1 T_1 + m_2 T_2) + 0.87(m_1 q_1 T_1 + m_2 q_2 T_2)}{m + 0.87 m_v}. \tag{7.73}$$

In the above expression $m = m_1 + m_2$ is the total mass and $m_v = m_1 q_1 + m_2 q_2$ is the total water vapor mass. Since there is no condensation, m_v remains constant. If the final specific humidity is q, then the water vapor mass is qm. It follows that

$$q = \frac{m_1 q_1 + m_2 q_2}{m} \tag{7.74}$$

i.e. the final q is the weighted average of q_1 and q_2. Using equation (7.74) we can rewrite equation (7.73) as

$$T = \frac{(m_1 T_1 + m_2 T_2) + 0.87(m_1 q_1 T_1 + m_2 q_2 T_2)}{m(1 + 0.87 q)}. \tag{7.75}$$

If we neglect all vapor terms the above equation becomes

$$T \approx \frac{m_1 T_1 + m_2 T_2}{m} \tag{7.76}$$

indicating that, approximately, the temperature of the mixture is also the weighted average of the initial temperatures. One can show that the potential temperature and the vapor pressure of the mixture will also, approximately, be the weighted averages of the initial values (see problem 7.1):

$$\theta \approx \frac{m_1 \theta_1 + m_2 \theta_2}{m} \tag{7.77}$$

$$e \approx \frac{m_1 e_1 + m_2 e_2}{m}. \tag{7.78}$$

Equations (7.76) and (7.78) indicate that in a vapor pressure–temperature diagram (figure 7.3) if the two masses correspond to points A_1 and A_2 then the mixture will correspond to the point A which will lie on the straight line connecting A_1 and A_2. From the similarity of triangles $AA_1'A_1$ and $A_2A_2'A$ it follows that point A will be found at a distance such that $A_1A/AA_2 = AA_1'/A_2A_2' = (T_1 - T)/(T - T_2) = m_2/m_1$ (here equation (7.76) was used). The thin solid lines in figure 7.3 represent the equal relative humidity curves corresponding to points A_1, A_2, and A. Because of their curvature it can easily be seen that the final relative humidity will be greater than the weighted average. This can easily be visualized if we consider the simple case where equal air masses have the same relative humidity, i.e. points A_1, A_2 lie on the same equal relative humidity curve (figure 7.4). Then point A will lie to the left of the curve indicating that the final relative humidity is greater than the weighted average, $(mr + mr)/2m = r$. We conclude that isobaric adiabatic mixing increases the relative humidity. This is a very interesting result because it implies that two air masses that

Figure 7.3
When two initially
unsaturated air masses
(A_1 at r_1 and A_2 at r_2)
mix isobarically, the final
product (A) may be
saturated.

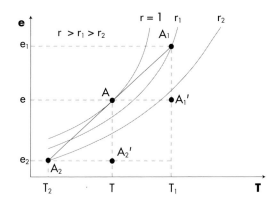

Figure 7.4
When two saturated
($r = 1$) air masses (A_1
and A_2) mix isobarically
the final product (A) is
supersaturated.

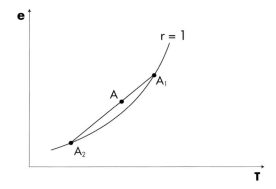

are initially unsaturated (but close to saturation) may result in a
mixture that is supersaturated. In this case condensation will occur. This mechanism can produce mixing fogs and other important
processes in the atmosphere.

7.3.2 Vertical mixing

Let us now look at vertical mixing occurring because of turbulent
and/or convective processes. Such a process is much more complicated than the previous one because now the vertical variation
of T and p come into the picture. However, we can simplify the
treatment of vertical mixing by (1) considering two isolated air
masses m_1 and m_2 at pressure levels p_1 and p_2 ($p_1 > p_2$) and temperatures T_1 and T_2, that are taken to a pressure level p where
they mix isobarically, and (2) redistributing the mixture in the
layer $\Delta p = p_1 - p_2$. The excursions to level p can be considered as
adiabatic ascent and descent (during which the specific humidities
q_1 and q_2 remain constant) resulting in new temperatures for the

air masses:

$$T_1' = T_1 \left(\frac{p}{p_1} \right)^k$$

$$T_2' = T_2 \left(\frac{p}{p_2} \right)^k.$$

Once at level p the air masses mix isobarically. From what we discussed in the previous section it follows that in this case

$$T \approx \frac{m_1 T_1' + m_2 T_2'}{m}$$

$$q \approx \frac{m_1 q_1 + m_2 q_2}{m} \tag{7.79}$$

and

$$\theta \approx \frac{m_1 \theta_1 + m_2 \theta_2}{m}$$

where θ_1 and θ_2 are the initial potential temperature values, which remain constant during the adiabatic ascent and descent. The subsequent redistribution of the mixture in the layer Δp will preserve θ assuming that it consists of adiabatic ascents and descents. It follows that when the layer is well mixed, θ will not vary with height. We can generalize equations (7.79) by considering mixing of n air masses and employing the general expression for the weighted average. For example,

$$\overline{\theta} = \frac{\int_0^m \theta dm}{m},$$

$$\overline{q} = \frac{\int_0^m q dm}{m},$$

$$\overline{w} = \frac{\int_0^m w dm}{m}.$$

Using the hydrostatic approximation we can express the above equations as

$$\overline{x} = \frac{\int_0^z x \rho dz}{\int_0^z \rho dz} = -\frac{\int_{p_1}^{p_2} x dp}{p_1 - p_2}$$

where x is θ, or q, or w.

The temperature variation with pressure can then be expressed according to

$$T = \overline{\theta} \left(\frac{p}{1000} \right)^k. \tag{7.80}$$

If in the initial distribution (before mixing) there is a level $(p, T <\overline{T})$, where $r = \overline{r}$, then this level is defined as the mixing condensation level.

Examples

(7.1)   The initial state of a parcel of air is $p = 1000$ mbar, $T = 30$ °C, $w = 14$ g kg^{-1}. Find (a) r, (b) T_{virt}, (c) T_{dew}, (d) T_w, (e) T_{ei}, (f) T_{LCL}, (g) z_{LCL}, (h) θ_{ep}, (i) T_{ep}, (j) θ_{wp}, (k) θ_{virt}.

(a)   From equation (6.17) we estimate that $e_{sw}(T) = 43.1$ mbar. From equation (7.3) we then find that $e(T) \approx 22.0$ mbar. Thus, $r \approx 0.51$, $w_{sw} \approx 28$ g kg^{-1}.

(b) $T_{virt} = (1+0.61q)T = \left(1 + 0.61\left(\frac{w}{1+w}\right)\right)T = 305.55$ K.

(c)   From equation (7.16) we obtain $T_{dew} = 291.75$ K.

(d)   From equation (7.22) we have

$$T_w = T + \frac{l_v}{c_{pd}}(w - w_{sw})$$

where $l_v(T = 303$ K$) = 2.43 \times 10^6$ J kg^{-1}. Now recall that w_{sw} in the above formula is the saturation mixing ratio at T_w. Making use of the approximations in equations (7.5) and (6.17) we have

$$T_w = T + \frac{l_v \epsilon}{c_{pd}p}\left[e(T) - 6.11 \exp\left(19.83 - \frac{5417}{T_w}\right)\right]$$

Numerical solution of the above equation gives $T_w \approx 295.4$ K.

(e)   From equation (7.23) for $c_w = 4218$ J kg^{-1} we find that $T_{ei} \approx 335$ K.

(f)   From equation (7.36) it follows that $T_{LCL} \approx 289$ K.

(g)   From equation (7.45) and assuming that the initial level is $z = 0$ we find that $z_{LCL} = 1.4$ km.

(h)   Using equation (7.63) with $T = 303$ K, $p = 1000$ mbar, $w = 0.014$ g kg^{-1}, $T_{LCL} = 289$ K, we obtain that $\theta_{ep} \approx 345$ K.

(i)   From equation (7.64) it follows (since $T = \theta$) that $T_{ep} = \theta_{ep} = 345$ K.

(j)   Considering that $w' = w_s (1000$ mbar, $T = \theta_{wp}) = (\epsilon/p)6.11 \exp(19.83 - (5417/\theta_{wp}))$, numerical solution of equation (7.65) yields $\theta_{wp} \approx 295$ K.

(k)   From equation (7.28) and since $p = 1000$ mbar it follows that $\theta_{virt} = T_{virt} = 305.55$ K.

Note that all estimated temperatures and potential temperatures are consistent with equation (7.66).

(7.2)   The temperature of saturated air at 1000 mbar is 10 °C as it begins to drop through radiation loss. As a result, some water vapor condenses and radiation fog forms. During the formation of fog 12×10^3 J kg^{-1} of heat is lost to the surroundings. Find the final temperature and the decrease in vapor pressure.

The process is non-adiabatic and thus non-isenthalpic. From equation (7.20) we have

$$dh = c_p dT + l_v dw$$

where $c_p = c_{pd} + w_t c_w$ and we have assumed that $dH/m_d \approx dh$. During the process the air remains in a state of saturation, thus w corresponds to saturation. Also, the process is isobaric, so $dh = \delta q$ and we can write the above equation as

$$\delta q = c_p dT + l_v dw_{sw}. \tag{7.81}$$

Using the approximate relationship $w_{sw} \approx \epsilon e_{sw}/p$, we find that $dw_{sw} \approx (\epsilon/p)de_{sw}$ or, using the C–C equation,

$$dw_{sw} = \frac{\epsilon}{p} \frac{l_v e_{sw}}{R_v T^2} dT. \tag{7.82}$$

By substituting (7.82) into (7.81) we obtain

$$\delta q = \left(c_p + \frac{\epsilon l_v^2 e_{sw}}{p R_v T^2} \right) dT.$$

In the above equation, c_p is to a good approximation independent of temperature. Also, for small temperature changes, l_v is nearly independent of temperature. The saturation vapor pressure, however, is dependent on T (via equation (6.17)). This makes the integration of the above equation very difficult. One way to overcome this difficulty is to assume that e_{sw}/T^2 is nearly constant for small temperature differences and then treat the differentials as differences. In this case, for $T = 10$ °C, $e_{sw} = 12.16$ mbar, $l_v(10$ °C$) = 2.4774 \times 10^6$ J kg^{-1}, $\delta q = -12 \times 10^3$ J kg^{-1}, and $c_p \approx c_{pd} = 1005$ J kg^{-1} K^{-1}, we find that $\Delta T = -5.3$ °C. Thus, the final temperature is 4.7 °C. Note that this result can be improved if we do not assume that $c_p \approx c_{pd}$. However, this will require the value of w_t which is not known. We could of course assume that $w_t \approx w_{sw}(10$ °C$)$ and consider $c_p \approx c_{pd} + w_{sw}c_w$. In this case, we find that the difference in the result is small (~ 0.066 °C).

In order to find the decrease in vapor pressure we combine equation (7.81) with $dw_{sw} \approx (\epsilon/p)de_{sw}$ to arrive at

$$\delta q = c_p dT + \frac{l_v \epsilon}{p} de_{sw}.$$

Then incorporating the C–C equation gives

$$\delta q = \left(\frac{c_p R_v T^2}{l_v e_{sw}} + \frac{l_v \epsilon}{p} \right) de_{sw}.$$

Figure 7.5
Diagram for example 7.3.
When two initially
saturated air masses mix
isobarically, the final
product is supersaturated
(T, e). In order for the
mixture to reach the
equilibrium state (T', e'),
the excess vapor will
condense. The amount of
vapor that condenses is
proportional to the
difference $e - e'$.

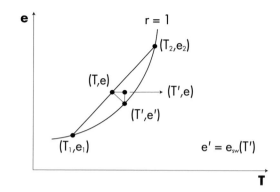

Here again if we treat the differentials as differences and use the values mentioned above we find that $\Delta e_{sw} = -4.326$ mbar.

(7.3) Two equal air masses, both at 1000 mbar, mix adiabatically. Their initial temperatures and mixing ratios are $T_1 = 293.8$ K, $w_1 = 16.3$ g kg^{-1}, $T_2 = 266.4$ K, $w_2 = 1.3$ g kg^{-1}. First show that the mixture is supersaturated. Then find the final temperature, the final mixing ratio, the amount of liquid water content per unit mass of air, and the amount of liquid water per unit volume of air.

As the two air masses mix isobarically, the initial temperature of the mixture will be (equation (7.75)) $T = 280.2$ K. Since $q_1 = 16.0386$ g kg^{-1} and $q_2 = 1.2983$ g kg^{-1} we find that, for the mixture, $q = 8.6685$ g kg^{-1} (equation (7.74)), $w = q/(1 - q) = 8.7443$ g kg^{-1}, and $e = wp/(\epsilon + w) = 13.86$ mbar (using the approximate equations (7.76) and (7.78) results in $T = 280.7$ K and $e = 14.1$ mbar).

For $T = 280.2$ K equation (6.17) yields $e_{sw} = 10.05$ mbar. This result indicates that initially the mixture is supersaturated $(e > e_{sw})$. Thus, immediately after mixing the excess vapor will condense so that the mixture reaches the equilibrium state (T', e'). This step, graphically shown in figure 7.5, will result in a decrease in vapor pressure and an increase in temperature (remember the mixing is adiabatic, so the heat gain due to condensation remains in the mixture). The increase in temperature is denoted by the line connecting points (T, e) and (T', e) and the decrease in water vapor pressure by the line connecting points (T', e') and (T', e). Since the process from (T, e) to (T', e') is isobaric and adiabatic it is isenthalpic. As such, equation

(7.21) applies:

$$(c_{pd} + w_t c_w)(T - T') = (w' - w)l_v$$

or

$$(c_{pd} + w_t c_w)(T - T') = (e' - e)l_v \frac{\epsilon}{p}$$

or

$$(c_{pd} + w_t c_w)(T - T') = \left[6.11 \exp\left(19.83 - \frac{5417}{T'} \right) - e \right] l_v \frac{\epsilon}{p}$$

where, assuming that the mixture is a closed system, $w_t = w = 8.7443$ g kg^{-1}. Numerical solution of the above equation, with $c_w = 4218$ J kg^{-1} K^{-1} and $l_v = 2.4774 \times 10^6$ J kg^{-1}, gives $T' \approx 282.8$ K. This corresponds to $e' = e_{sw}(T') = 12$ mbar and hence to $w' = 7.55$ g kg^{-1}.

The amount of liquid water produced per unit mass of air is given by the difference $q - q' \approx w - w' \approx \frac{\epsilon}{p}(e - e') \approx 1.16$ g kg^{-1}. This amount must be equal to the decrease of water vapor per unit mass, $-dm_v$, which is proportional to the decrease in water vapor pressure, $-de_{sw}$. From the ideal gas law we know that the mass of water vapor per unit volume at temperature T is $e_{sw}/R_v T$ and its variation with temperature is

$$de_{sw}/dT = de_{sw}/R_v T - (e_{sw}/R_v T^2)dT.$$

Using the C–C equation we can verify that the second term on the right-hand side of the above equation is much smaller than the first term. As such, the amount of water vapor lost is approximately equal to $de_{sw}/R_v T$. It follows, that the amount of liquid water *per volume* is equal to $de_{sw}/R_v T$, or equal to

$$\frac{e - e'}{R_v T} \approx 1.44 \text{ g m}^{-3}.$$

(7.4) The following problem (Salby (1996)) is a classic example, demonstrating the effect of the downslope wind called Chinook wind (Chinook in American Indian means "snow-eater"). Moist air from the Pacific is forced to rise over the continental divide. On the west side, the surface lies at 800 mbar where the temperature is 293 K and the mixing ratio is 15 g kg^{-1}. If the summit lies at 600 mbar and assuming that any condensation products fall out, find the surface temperature and relative humidity at a place on the leeward side at 830 mbar.

For $T = 293$ K equation (6.17) gives $e_{sw} = 23.4$ mbar. From equation (7.4) we then find that the saturation mixing ratio at state (T, p, w) : (293 K, 800 mbar, 15 g kg^{-1}) is

18.74 g kg^{-1}. Since $w < w_s$, the parcel is initially unsaturated. As the air begins to rise over the mountain it cools and its relative humidity increases. From equation (7.36) we find that, for $r = 15/18.74 = 0.8$, $T_{LCL} = 288.7$ K.

The ascent to the level where the air first becomes saturated is described by Poisson's equation

$$p_{LCL} = p \left(\frac{T_{LCL}}{T} \right)^{1/k}$$

where $k = k_d(1 - 0.26w) = 0.2849$. It follows that $p_{LCL} \approx 760$ mbar. At this point the air is saturated. As it continues to rise condensation takes place and all condensation products fall out. As such, the ascent above p_{LCL} is described by a pseudo-adiabatic process. Therefore, θ_{ep} from 760 to 600 mbar is conserved. From equation (7.63) we find, for $T = T_{LCL} = 288.7$ K, $p = p_{LCL} = 760$ mbar, and $w = w_{sw}(T_{LCL}) = 15$ g kg^{-1}, that $\theta_{ep} \approx 359$ K. Evaluating the same equation at $p' = 600$ mbar gives

$$359 = T' \left(\frac{1000}{p'} \right)^{0.285(1-0.28w')}$$
$$\times \exp(w'(1 + 0.81w') \left(\frac{3376}{T_{LCL}} - 2.54 \right) \quad (7.83)$$

where T', p', w' is the state of the air at the summit. In the above equation we know p' but T' and w' are unknowns. However, w' can be expressed as a function of T' alone by combining equations (7.5) and (6.17):

$$w' = w_{sw}(T') \approx \frac{\epsilon e_{sw}(T')}{p'} = \frac{6.11\epsilon}{p'} \exp \left(19.83 - \frac{5417}{T'} \right).$$

Thus, in effect equation (7.83) has only one unknown, T'. Solving this equation numerically gives $T' \approx 279$ K. It follows that at the summit $w' \approx 9.6$ g kg^{-1}. From that point on the air descends adiabatically with the mixing ratio remaining constant (9.6 g kg^{-1}). From Poisson's equation we then find that at 830 mbar, $T \approx 306$ K. At this temperature $w_{sw} \approx 38.5$ g kg^{-1} and at $p = 830$ mbar the relative humidity is $r = 0.25$. Thus the air on the leeward side is warmer and drier than that on the west side. Note that we assume that the parcel from 760 to 600 mbar remains saturated with respect to *liquid* water. This is a safe assumption because $T_{LCL} = 288.7$ K and the vertical distance between 760 and 600 mbar in a standard atmosphere is about 1.5 km. So, we would expect that the temperature at the summit would be above 0 °C.

Problems

(7.1) Show that when two air masses mix adiabatically and iso-barically without condensation taking place, the final potential temperature and final vapor pressure are the weighted averages of the initial values (make necessary assumptions).

(7.2) Show that

$$\frac{p_{\text{LCL}}}{e_{\text{sw}}(T_{\text{LCL}})} = \frac{p}{e_{\text{sw}}(T_{\text{dew}})}.$$

Then use equation (6.17) to derive a relationship between $T_{\text{LCL}}, T_{\text{dew}}, p$, and p_{LCL}.

(7.3) An air mass has a temperature of 20 °C at 970 mbar pressure with a mixing ratio of 5 g kg^{-1}. After some time this mass has acquired a temperature of 10 °C and a pressure of 750 mbar. Assume that no condensation or mixing between the air mass and the environment takes place and calculate the initial and final values of vapor pressure, relative humidity and dew point temperature. (7.74 mbar, 5.98 mbar, 0.33, 0.49, 276.5 K, 272.8 K)

(7.4) Show that evaporation of dm grams of water into m_{d} grams of dry air (under constant temperature) requires an amount of heat absorbed given by

$$\delta Q = m_{\text{d}} l_{\text{v}} dw.$$

(7.5) A Chinook wind blowing at 800 mbar has a temperature of 38 °C and a mixing ratio of 4 g kg^{-1}. Is this the same as the air at 1000 mbar on the windward side of the mountains having a temperature of 294.5 K and a mixing ratio of 10 g kg^{-1} or is it the same as the air at 800 mbar with 278 K and 5 g kg^{-1}? (It is the same as the air at 1000 mbar)

(7.6) Show that if an isothermal layer of temperature T_0 contained between isobars p_1 and p_2 mixes vertically, the final temperature distribution is given by

$$T(p) = [T_0/(1-k)] \left(p_1^{1-k} - p_2^{1-k}\right) p^k/(p_1 - p_2).$$

(7.7) What will be the lowest possible temperature to which very dry air at $T = 40$ °C can be cooled by evaporation? Consider l_{v} constant and equal to 2.4×10^6 J kg^{-1}. (~ 288 K)

(7.8) When it is cold outside our breath creates mixing clouds. Is it possible to create a mixing cloud when the temperature of our breath is less than the air temperature? Assume that the temperature of breath is 30 °C. (Hint: Plot e_{sw} as a function of T using equation (6.17) and make assumptions about this curve for high temperatures. Then use the theory discussed in section 7.3.1.)

(7.9) On a summer day in Milwaukee the high temperature is pre-
dicted to reach 35 °C with dew point around 28 °C. What
will be the lowest temperature to which the air can be cooled
by evaporation in this case? Does the answer make sense?
Elaborate on the difference between this answer and the
answer in problem 7.7. Consider l_v constant and equal to
2.4×10^6 J kg^{-1}. ($\sim$ 303 K)

(7.10) Moist air moves over land where the temperature is 20 °C
and the relative humidity is 70%. Through contact with
the ground the air cools. At what temperature will fog
form? Consider l_v constant and equal to 2.45×10^6 J kg^{-1}.
(287.5 K)

(7.11) The main cabin in an airplane is maintained at a pressure
of 900 mbar and at a temperature of 25 °C. If it suddenly
decompresses adiabatically to a pressure of 500 mbar, what
should the relative humidity be to avoid cloud formation in
the cabin? (6.29%)

(7.12) Outside air has a temperature of -15 °C and a relative
humidity of 0.6. If the air indoors is 25 °C, what is the
relative humidity inside the room (assuming that the air
inside is only heated, not humidified)? If the room has a
volume of 100 m^3 how much water vapor must be supplied
so that the relative humidity rises to 50% and what will then
be the total mixing ratio? If the temperature change due to
evaporating water is neglected, what amount of heat must
be added for this to happen? Assume l_v constant and equal
to 2.5×10^6 J kg^{-1}. (3.6%, 1.08 kg, 8.97 g kg^{-1}, 2.7×10^6 J)

(7.13) Which of the following quantities are conserved in (a)
reversible adiabatic unsaturated transformation, (b) reversible adiabatic saturated transformation, (c) adiabatic unsaturated isobaric transformation, (d) adiabatic saturated
isobaric transformation?

$w, q, w_t, q_t, e, e_{sw}, w_{sw}, r, p_{LCL}, T_{LCL}, T_w, \theta, \theta_{virt}, \theta_e, \theta_w, s, h.$

(7.14) A refrigerator of volume 2 m^3 is initially filled with air at
temperature 303 K and with a relative humidity of 50%.
At what temperature condensation will form on the walls?
Assuming that the desired final temperature is 275 K, how
much water vapor must condense to achieve the final temperature? What is the total amount of heat that will be
given away to the surroundings during the whole process?
Assume $l_v = 2.5 \times 10^6$ J kg^{-1}, $c_w = 4218$ J kg^{-1}K^{-1},
$\rho_d = 1.293$ kg m^{-3}. ($\sim$ 19 °C, 19.8 g, 117 368 J)

(7.15) If the temperature inside a room is 25 °C and the temperature outside is -10 °C calculate the maximum relative
humidity that can be accommodated inside the room without windows fogging when (a) the interior part of the win-

dows is not thermally isolated from the exterior part, and
(b) it is thermally isolated. Assume $l_v = 2.5 \times 10^6$ J kg^{-1},
$c_w = 4218$ J kg^{-1}K^{-1}. $(0.32, 1.0)$

(7.16) If the wet-bulb temperature of a parcel at 1000 mbar is
15.6 °C and its mixing ratio is 6 g kg^{-1}, find the relative
humidity of the parcel after it is lifted dry adiabatically to
900 mbar. Assume $l_v = 2.47 \times 10^6$ J kg^{-1}. (36.14%)

(7.17) Raindrops at 10 °C evaporate into air of temperature 20 °C.
If the saturated mixing ratio of air at 10 °C is 8 g kg^{-1}, what
is the mixing ratio of the environmental air? Make assumptions and consider $l_v = 2.47 \times 10^6$ J kg^{-1}. $(3.93$ g kg$^{-1})$

(7.18) If water vapor comprises 1% of the volume of air (i.e. it
accounts for 1% of the molecules) what is the virtual temperature correction? $(T_{virt} = 1.0038 \, T)$

(7.19) If all the water vapor in a column of air extending from the
surface to the troposphere were to condense and fall to the
surface, it would occupy the lower d_w of the height of the
column. In this case, we can write that

$$\rho_w d_w = \int_0^\infty \rho_v dz$$

where ρ_w and ρ_v are the densities of liquid water and water
vapor, respectively. Show that the *upper* limit for d_w is given
by

$$d_w \approx \frac{T_0 e_{s0}}{\rho_w l_v \Gamma}$$

where Γ is the lapse rate, which is assumed to be constant,
T_0 is the surface temperature, and e_{s0} is the saturation vapor pressure with respect to water at T_0. Get an estimate
of d_w by assuming realistic values for T_0 and Γ. What do
you think of the estimated d_w? Is it realistic? [d_w is often
called precipitable water.]

(7.20) Show that the mixing ratio of a parcel of air is given exactly
by

$$w = \frac{r w_s}{1 + (1 - r)\frac{w_s}{\epsilon}}.$$

(7.21) In *Meteorology* Book I, Aristotle says: "Both frost and dew
are found when the sky is clear and there is no wind." What
do you think? Is he right or wrong? Explain why.

CHAPTER EIGHT

Vertical stability in the atmosphere

Over large scales the atmosphere is very nearly hydrostatic, which means that the pressure gradient force balances the force of gravity. Because the net force in the vertical is zero, the atmosphere over those scales exhibits slow vertical motions of constant speed, i.e. there is no upward acceleration of air. However, on smaller scales hydrostatic equilibrium may be invalidated. In this case, accelerated motion leads to convection which embraces many phenomena observed in the atmosphere, from the structure of planetary boundary layers to the dynamics of hurricanes. In this chapter we will investigate the conditions that allow such accelerating motions. In particular, we will examine the fate of a parcel at equilibrium with its environment when it is subjected to a small perturbation. We will make the following assumptions: (1) the environment is in hydrostatic equilibrium; (2) the parcel does not mix with its surroundings; (3) the parcel's movement does not disturb the environment; (4) the process is adiabatic; and (5) at a given level the pressure of the environment and the pressure of the parcel are equal. In problem 8.1 you will be asked to elaborate on the validity of the above assumptions.

8.1 The equation of motion for a parcel

Because the environment is in hydrostatic equilibrium the following equation holds:

$$\frac{dp}{dz} = -\rho g.$$

For the parcel, the above equation is not applicable because as the parcel is displaced (upwards or downwards) it has some acceleration (d^2z/dt^2). Assuming that gravity and the pressure gradient force are acting on the parcel, Newton's second law dictates that per unit volume (primes identify the parcel)

$$\rho' \frac{d^2 z}{dt^2} = -\rho' g - \frac{dp'}{dz}$$

or

$$\ddot{z} = -g - a' \frac{dp'}{dz}.$$

Because of assumption (5) above, $dp'/dz = dp/dz$. Thus, the above equation can be written as

$$\ddot{z} = -g - a'\left(-\frac{g}{a}\right)$$

or

$$\ddot{z} = g\left(\frac{a' - a}{a}\right)$$

or

$$\ddot{z} = g\left(\frac{\rho - \rho'}{\rho'}\right). \tag{8.1}$$

The right-hand side gives the force per unit mass acting on the parcel due to the combination of gravity and pressure gradient and is called the *buoyancy* of the parcel. Using the definition of virtual temperature and assumption (5), we can write the ideal gas law for both environment and parcel as:

$$p = \rho R_{\mathrm{d}} T_{\mathrm{virt}}$$

and

$$p = \rho' R_{\mathrm{d}} T'_{\mathrm{virt}}. \tag{8.2}$$

The reason for adopting these expressions rather than $p = \rho R T$ and $p = \rho' R' T'$, is that this way we deal with T_{virt} and T'_{virt} rather than R, R', T, and T'. This makes the analysis more straightforward.

Combining equations (8.1) and (8.2) yields

$$\ddot{z} = g\left(\frac{T'_{\mathrm{virt}} - T_{\mathrm{virt}}}{T_{\mathrm{virt}}}\right). \tag{8.3}$$

What we are interested in here is to investigate the effect on the motion, described by equation (8.3), of small displacements ($z \ll 1$) from an original equilibrium level. If for simplicity, we take this level to be the $z = 0$ level where the temperature is $T_{\mathrm{virt},0}$ and express both T_{virt} and T'_{virt} in terms of Taylor's series we get

$$T_{\mathrm{virt}} = T_{\mathrm{virt},0} + \frac{dT_{\mathrm{virt}}}{dz} z + \frac{1}{2} \frac{d^2 T_{\mathrm{virt}}}{dz^2} z^2 + \cdots$$

and

$$T'_{\mathrm{virt}} = T_{\mathrm{virt},0} + \frac{dT'_{\mathrm{virt}}}{dz} + \frac{1}{2} \frac{d^2 T'_{\mathrm{virt}}}{dz^2} z^2 + \cdots$$

If we now neglect terms of order higher than the first and define the environmental virtual temperature lapse rate as $-dT_{\text{virt}}/dz = \Gamma_{\text{virt}}$ and the parcel's virtual temperature lapse rate as $-dT'_{\text{virt}}/dz = \Gamma'_{\text{virt}}$ we arrive at

$$\ddot{z} = \frac{g(\Gamma_{\text{virt}} - \Gamma'_{\text{virt}})z}{T_{\text{virt},0} - \Gamma_{\text{virt}}z}. \tag{8.4}$$

In the above equation we can manipulate the term $1/(T_{\text{virt},0} - \Gamma_{\text{virt}}z)$ to arrive at

$$\frac{1}{T_{\text{virt},0} - \Gamma_{\text{virt}}z} = \frac{1}{T_{\text{virt},0}} \frac{1}{1 - \frac{\Gamma_{\text{virt}}z}{T_{\text{virt},0}}}$$

$$\approx \frac{1}{T_{\text{virt},0}} \left(1 + \frac{\Gamma_{\text{virt}}z}{T_{\text{virt},0}}\right) \quad \left(\text{because } \frac{\Gamma_{\text{virt}}z}{T_{\text{virt},0}} \ll 1\right).$$

Then equation (8.4) becomes

$$\ddot{z} = \frac{g}{T_{\text{virt},0}} \left(1 + \frac{\Gamma_{\text{virt}}z}{T_{\text{virt},0}}\right) (\Gamma_{\text{virt}} - \Gamma'_{\text{virt}})z$$

or by eliminating terms involving z^2

$$\ddot{z} = \frac{g}{T_{\text{virt},0}} (\Gamma_{\text{virt}} - \Gamma'_{\text{virt}})z$$

or

$$\ddot{z} + \frac{g}{T_{\text{virt},0}} (\Gamma'_{\text{virt}} - \Gamma_{\text{virt}})z = 0. \tag{8.5}$$

8.2 Stability analysis and conditions

The solution of the differential equation (8.5) depends on the constants. Three possibilities exist.

1 $\Gamma'_{\text{virt}} - \Gamma_{\text{virt}} > 0$

In this case equation (8.5) takes the form $\ddot{z} + \lambda^2 z = 0$ and has the solution

$$z(t) = A \sin \lambda t + B \cos \lambda t$$

where the oscillatory components are characterized by

$$\lambda = \sqrt{\frac{g}{T_{\text{virt},0}} (\Gamma'_{\text{virt}} - \Gamma_{\text{virt}})} > 0.$$

This is the so-called Brunt–Väisälä frequency. Since we assumed that the initial level is $z = 0$, it follows that in this case $B = 0$ and as such $z(t) = A \sin \lambda t$. This indicates that the parcel will oscillate in time about its original position with a period $\tau = 2\pi/\lambda$. This represents a stable case as the parcel does not leave the original level.

2 $\Gamma'_{\text{virt}} - \Gamma_{\text{virt}} < 0$

In this case equation (8.5) takes the form $\ddot{z} - \lambda^2 z = 0$ and has the solution

$$z(t) = Ae^{\lambda t} + Be^{-\lambda t}$$

where now

$$\lambda = \sqrt{\frac{g}{T_{\text{virt},0}}(\Gamma_{\text{virt}} - \Gamma'_{\text{virt}})} > 0.$$

Since at $t = 0$, $z(0) = 0$ it follows that $A + B = 0$. This indicates that $A = -B \neq 0$ (the possibility of $A = B = 0$ leads to the trivial solution $z(t) = 0$ which we ignore). Since $A \neq 0$ it then follows that as $t \to \infty$, a parcel's displacement grows exponentially. Note that since for $t \to \infty$, $dz/dt = A\lambda e^{\lambda t}$, the parcel's motion is an accelerating motion. This is the unstable situation where the parcel leaves the original level and never returns.

3 $\Gamma'_{\text{virt}} - \Gamma_{\text{virt}} = 0$

In this case equation (8.5) becomes $\ddot{z} = 0$ and has the linear solution

$$z(t) = At + B$$

indicating that the displacement grows linearly with time. The parcel's motion is now one of a constant speed ($dz/dt = A$). This is the neutral case where the parcel leaves the original level and never returns. Thus, the only difference between neutral and unstable situations is whether or not the motion is an accelerating one.

From equation the physical meaning of the above is that as long as the parcel's lapse rate is greater than that of the environment and the parcel is forced upwards (downwards), it will become colder (warmer) than the environment and it will sink (rise) back to the original level. If the parcel's lapse rate is less than that of the environment and the parcel is forced upwards (downwards) it will become warmer (colder) than the environment and it will continue to rise (sink) thus moving away from the original level. The difference between the two lapse rates is the net force acting on the parcel. If the net force is zero the motion is one of a constant speed. Otherwise the net force acts in the direction of the initial impulse if $\Gamma'_{\text{virt}} - \Gamma_{\text{virt}} < 0$ or in the opposite direction if $\Gamma'_{\text{virt}} - \Gamma_{\text{virt}} > 0$.

From equation (7.9) we have that for an *unsaturated* parcel $T'_{\text{virt}} = (1 + 0.61w')T'$, where w' is the initial mixing ratio, which because of assumption (2) remains constant. Thus,

$$\frac{dT'_{\text{virt}}}{dz} = (1 + 0.61w')\frac{dT'}{dz}$$

or

$$\Gamma'_{\text{virt}} = (1 + 0.61w')\Gamma_{\text{m}}$$

or using equation (7.29)

$$\Gamma'_{\text{virt}} = (1 + 0.61w')(1 - 0.87w')\Gamma_{\text{d}}$$

or

$$\Gamma'_{\text{virt}} \approx (1 - 0.26w')\Gamma_{\text{d}}.$$

The term $0.26w'$ in the above relation is rather small and therefore we may neglect it. Thus, for *unsaturated* parcels

$$\Gamma'_{\text{virt}} \approx \Gamma_{\text{d}}. \tag{8.6}$$

Note that for the environment, w is not constant but a function of z. As such, for the environment we have

$$\frac{dT_{\text{virt}}}{dz} = (1 + 0.61w)\frac{dT}{dz} + 0.61T\frac{dw}{dz}$$

or

$$\Gamma_{\text{virt}} = (1 + 0.61w)\Gamma - 0.61T\frac{dw}{dz}. \tag{8.7}$$

The second term on the right-hand side of equation (8.7) is not negligible and therefore we cannot arrive at a relation similar to equation (8.6) for Γ_{virt} and Γ (the environmental lapse rate). This is why we have to use the virtual temperature in our stability analysis rather than the actual temperature.

We can then state the condition for stability for an *unsaturated* parcel as follows:

$$\begin{array}{lll} \text{if} & \Gamma_{\text{virt}} > \Gamma_{\text{d}} & \text{the parcel is unstable,} \\ \text{if} & \Gamma_{\text{virt}} = \Gamma_{\text{d}} & \text{the parcel is neutral,} \\ \text{if} & \Gamma_{\text{virt}} < \Gamma_{\text{d}} & \text{the parcel is stable.} \end{array} \tag{8.8}$$

For *saturated* parcels the mixing ratio decreases with height. Proceeding as above we have

$$T'_{\text{virt}} = (1 + 0.61w')T'$$

or

$$\frac{dT'_{\text{virt}}}{dz} = (1 + 0.61w')\frac{dT'}{dz} + 0.61T'\frac{dw'}{dz}$$

or

$$\Gamma'_{\text{virt}} = (1 + 0.61w')\Gamma_{\text{s}} - 0.61T'\frac{dw'}{dz}.$$

In this case the second term of the right-hand side is much smaller than the first term. Thus, we can approximate the above equation as

$$\Gamma'_{\text{virt}} \approx \Gamma_{\text{s}}.$$

Figure 8.1
Relative position of fundamental lines and regions of stability.

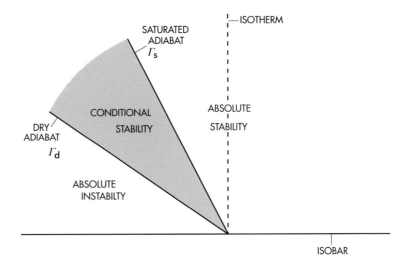

Accordingly, the condition for stability for a *saturated* parcel can be stated as

$$
\begin{array}{lll}
\text{if} & \Gamma_{\text{virt}} > \Gamma_{\text{s}} & \text{the parcel is unstable,} \\
\text{if} & \Gamma_{\text{virt}} = \Gamma_{\text{s}} & \text{the parcel is neutral,} \\
\text{if} & \Gamma_{\text{virt}} < \Gamma_{\text{s}} & \text{the parcel is stable.}
\end{array}
\tag{8.9}
$$

Since $\Gamma_{\text{d}} = 9.8\ °C\ \text{km}^{-1}$ and $\Gamma_{\text{s}} \approx 5\ °C\ \text{km}^{-1}$, conditions (8.8) and (8.9) can be combined in one as follows:

$$
\begin{array}{lll}
\text{if} & \Gamma_{\text{virt}} > \Gamma_{\text{d}} & \text{the parcel is absolutely unstable,} \\
\text{if} & \Gamma_{\text{s}} < \Gamma_{\text{virt}} < \Gamma_{\text{d}} & \text{the parcel is conditionally unstable,} \\
\text{if} & \Gamma_{\text{virt}} < \Gamma_{\text{s}} & \text{the parcel is absolutely stable.}
\end{array}
\tag{8.10}
$$

The word "absolutely" indicates that the stability criterion holds for any type (saturated or unsaturated) of parcel. The term "conditionally unstable" means that the parcel is stable if it is unsaturated and unstable if it is saturated (see figure 8.1).

Now recall that for unsaturated environments equation (7.28) applies:

$$
\theta_{\text{virt}} = T_{\text{virt}} \left(\frac{1000}{p} \right)^{k_{\text{d}}} .
$$

Logarithmic differentiation of the above equation gives

$$
\frac{1}{\theta_{\text{virt}}} \frac{d\theta_{\text{virt}}}{dz} = \frac{1}{T_{\text{virt}}} \frac{dT_{\text{virt}}}{dz} - \frac{k_{\text{d}}}{p} \frac{dp}{dz}
$$

or

$$
\frac{1}{\theta_{\text{virt}}} \frac{d\theta_{\text{virt}}}{dz} = -\frac{1}{T_{\text{virt}}} \Gamma_{\text{virt}} - \frac{k_{\text{d}}}{p} \left(-\frac{p}{R_{\text{d}} T_{\text{virt}}} g \right)
$$

or

$$\frac{1}{\theta_{\text{virt}}} \frac{d\theta_{\text{virt}}}{dz} = -\frac{\Gamma_{\text{virt}}}{T_{\text{virt}}} + \frac{1}{T_{\text{virt}}} \left(\frac{g}{c_{pd}} \right)$$

or

$$\frac{1}{\theta_{\text{virt}}} \frac{d\theta_{\text{virt}}}{dz} = \frac{1}{T_{\text{virt}}} \left(\Gamma_{\text{d}} - \Gamma_{\text{virt}} \right). \tag{8.11}$$

Combining (8.8) with (8.11) yields an alternative way to express stability conditions for unsaturated parcels:

$$\text{if} \quad \frac{d\theta_{\text{virt}}}{dz} > 0 \quad \text{the parcel is stable,}$$

$$\text{if} \quad \frac{d\theta_{\text{virt}}}{dz} = 0 \quad \text{the parcel is neutral,} \tag{8.12}$$

$$\text{if} \quad \frac{d\theta_{\text{virt}}}{dz} < 0 \quad \text{the parcel is unstable.}$$

For a saturated parcel similar conditions can be applied if we substitute θ_{virt} by θ_{e} (or in practice θ_{ep}) which is invariant along saturated adiabats (or pseudoadiabats)

$$\text{if} \quad \frac{d\theta_{\text{e}}}{dz} > 0 \quad \text{the parcel is stable,}$$

$$\text{if} \quad \frac{d\theta_{\text{e}}}{dz} = 0 \quad \text{the parcel is neutral,} \tag{8.13}$$

$$\text{if} \quad \frac{d\theta_{\text{e}}}{dz} < 0 \quad \text{the parcel is unstable.}$$

The above set of criteria applies to the stability of a parcel as it rises in a motionless layer. If this is not true (for example, when entire layers are lifted or lowered), then the stability of the parcels may be affected. Here is why.

Let us consider a stable layer in hydrostatic equilibrium whose bottom has a pressure p_1 and whose top has a pressure p_2. Let us further assume that the difference in pressure, Δp, between top and bottom remains constant during the lifting or sinking. If the whole process is an unsaturated adiabatic process, then equation (8.11) applies. Since in this process θ_{virt} is conserved then the difference in θ_{virt} between top and bottom is conserved. It follows that

$$\frac{1}{\theta_{\text{virt}}} \frac{d\theta_{\text{virt}}}{dz} = \frac{1}{\theta_{\text{virt}}} \frac{d\theta_{\text{virt}}}{dp} (-\rho g) = \frac{1}{T_{\text{virt}}} (\Gamma_{\text{d}} - \Gamma_{\text{virt}})$$

or

$$\frac{1}{\theta_{\text{virt}}} \frac{d\theta_{\text{virt}}}{dp} = -\frac{R_{\text{d}}}{pg} (\Gamma_{\text{d}} - \Gamma_{\text{virt}}) = \text{constant}$$

or

$$R_{\text{d}} (\Gamma_{\text{d}} - \Gamma_{\text{virt}}) = \text{constant} \times g \times p.$$

During lifting pressure decreases, and according to the above equation $\Gamma_{\text{virt}} \to \Gamma_{\text{d}}$. Since the layer is initially stable this tendency will cause the layer's lapse rate to approach the region of instability (recall figure 8.1). During sinking the opposite is true. As such, lifting of a layer tends always to destabilize the layer and sinking of a layer tends always to stabilize the layer.

If the lifting process is strong enough to cause the layer to saturate, then the picture is quite different. In this case the stability changes depend on the way saturation is reached, which in turn depends on the vertical distribution of moisture and the vertical structure of temperature among other factors. It can be shown that in this case the stability criteria for the layer translate to the following (a graphical demonstration of these criteria will be presented in Chapter 9):

$$
\begin{array}{lll}
\text{if} & \dfrac{d\theta_{\text{e}}}{dz} > 0 & \text{the saturated layer will be stable with} \\
& & \text{respect to saturated parcel processes,} \\[2mm]
\text{if} & \dfrac{d\theta_{\text{e}}}{dz} = 0 & \text{the saturated layer will be neutral with} \\
& & \text{respect to saturated parcel processes,} \\[2mm]
\text{if} & \dfrac{d\theta_{\text{e}}}{dz} < 0 & \text{the saturated layer will be unstable with} \\
& & \text{respect to saturated parcel processes.}
\end{array}
\qquad (8.14)
$$

where $d\theta_{\text{e}}/dz$ refers to the initial profile of θ_{e} (or in practice θ_{ep}) in the layer, which initially is not saturated. The above criteria are often referred to as *convective* stability, neutrality, and instability, respectively. Note that conditions (8.14) refer to the stratification of a layer initially unsaturated which is lifted *en masse* in the atmosphere while conditions (8.13) refer to the environment in which a saturated parcel rises (or sinks).

8.3 Other factors affecting stability

In our initial assumptions we accepted that the parcel does not mix with the environment, that the process is adiabatic, and that the environment is not disturbed (i.e. there are no compensating vertical motions in the environment as a parcel or a layer rises in it). When these assumptions are violated they affect the temperature lapse rate of the parcel and so the stability criteria derived previously do not apply. A detail treatment of these subjects is beyond the scope of this book. Interested readers can consult Iribarne and Godson (1973).

Examples

(8.1)   The table below provides pressure, temperature, and dew point temperature measurements at various pressure levels. (1) Investigate the stability conditions of parcels rising in each layer. (2) If each layer were lifted *en masse* until it became saturated would it become convectively stable, neutral, or unstable? (Assume that l_v is independent of T and equal to 2.45×10^6 J kg^{-1}.)

p (mbar)	T (°C)	T_{dew} (°C)
1000	30	23
950	27	21
900	23	20
850	20	20
800	18	10
750	15	5
700	10	2
650	5	0

First, we need to establish whether or not at a given level the parcels are saturated. The temperature is at all levels greater than the dew point temperature except at 850 mbar. Accordingly, for the stability of the parcels, conditions (8.12) apply for all layers except for the layer 850–800 mbar. For the layer 850–800 mbar, conditions (8.13) apply. For the convective stability of each layer conditions (8.14) apply. Thus, in order to answer the questions we need to calculate θ_{virt} and θ_e at each level. In order to estimate θ_{virt} and θ_e (recall equations (7.28) and (7.63)) we need to find at each level w and T_{LCL}. The mixing ratio can be estimated by combining equations (7.3), (7.7), (7.16), and (6.17). T_{LCL} can be estimated from equation (7.36). After plenty of calculations we arrive at the following table; here e_s is in mbar.

	p	T	T_{dew}	w	T_{virt}	θ_{virt}	θ_e	r	e_s	T_{LCL}
	1000	303	296	0.0182	306.4	306.4	357.4	0.66	43.1	294.3
layer 1										
	950	300	294	0.0169	303.0	307.5	355.1	0.70	36.0	292.6
layer 2										
	900	296	293	0.0166	299.0	308.1	355.0	0.83	28.2	292.3
layer 3										
	850	293	293	0.0176	296.1	310.2	360.2	1.00	23.4	293.0
layer 4										
	800	291	283	0.0099	292.8	312.1	340.1	0.61	20.6	281.3
layer 5										
	750	288	278	0.0149	290.6	315.5	361.7	0.51	17.0	276.0
layer 6										
	700	283	275	0.0063	284.1	314.6	333.3	0.58	12.2	273.4
layer 7										
	650	278	273	0.0058	279.0	315.6	332.9	0.70	8.6	272.0

(1)   According to the results in the above table and conditions (8.12) and (8.13), we find that for parcels rising in each of the layers, layer 1 is stable $(d\theta_{\text{virt}}/dz > 0)$, layer 2 is stable $(d\theta_{\text{virt}}/dz > 0)$, layer 3 is stable $(d\theta_{\text{virt}}/dz > 0)$, layer 4 is unstable $(d\theta_{\text{e}}/dz < 0)$, layer 5 is stable $(d\theta_{\text{virt}}/dz > 0)$, layer six is unstable $(d\theta_{\text{virt}}/dz < 0)$ and layer 7 is stable $(d\theta_{\text{virt}}/dz > 0)$.

(2)   Similarly, we find that layer 1 is convectively unstable $(d\theta_{\text{e}}/dz < 0)$, layer 2 is convectively neutral $(d\theta_{\text{e}}/dz \approx 0)$, layer 3 is convectively stable $(d\theta_{\text{e}}/dz > 0)$, layer 4 is convectively unstable, $(d\theta_{\text{e}}/dz < 0)$, layer 5 is convectively stable $(d\theta_{\text{e}}/dz > 0)$, layer 6 is convectively unstable $(d\theta_{\text{e}}/dz < 0)$ and layer 7 is approximately convectively neutral $(d\theta_{\text{e}}/dz \approx 0)$.

(8.2)   In an atmospheric layer the virtual temperature is constant and equal to 10 °C. If an initial fluctuation causes the parcel to rise adiabatically, calculate the energy per unit mass that has to be given to the parcel in order for it to rise 1 km above the bottom of the layer before it begins to sink.

When a parcel rises in the atmosphere, a certain amount of work is performed by or against the buoyancy force depending on whether the motion is along or against the direction of the buoyancy force. If the buoyancy force is directed downwards (negative buoyancy) a certain amount of work is done against buoyancy and if the buoyancy force is directed upwards (positive buoyancy) a certain amount of work is done by the buoyancy force. In the above problem since the layer is isothermal the layer is stable. Thus, the buoyancy on the parcel is negative and a parcel forced to rise will have to return to the initial level. But before it begins its return, it will reach a maximum height which depends on how strong the initial impulse was. Recall from Chapter 4 that

$$\delta W = F dz = ma dz = m\ddot{z} dz.$$

Thus, the work done, W, when a parcel is forced to rise from a level i to a level f is

$$W = \int_i^f m\ddot{z} dz. \tag{8.15}$$

Using equation (8.3) the above equation becomes

$$W = \int_i^f mg \left(\frac{T'_{\text{virt}}(z) - T_{\text{virt}}(z)}{T_{\text{virt}}(z)} \right) dz \tag{8.16}$$

or

$$W = gm \int_i^f \frac{T'_{\text{virt}}(z)}{T_{\text{virt}}(z)} dz - gm \int_i^f dz$$

or, assuming that the ascent is dry adiabatic,

$$W = gm \int_i^f \frac{(T_{\text{virt},0} - \Gamma_d z)}{T_{\text{virt}}(z)} dz - gm(f - i).$$

Since $T_{\text{virt}}(z) = T_{\text{virt},0}$ and assuming that the bottom of the layer is at $z = 0$ (i.e $i = 0$ m and $f = 1000$ m), we find that the work done per unit mass is equal to -169.7 J kg^{-1}, i.e. work is done against the buoyancy force. Since the ascent is assumed to be adiabatic, $\delta Q = 0$. Then from the first law we obtain

$$du = 169.7 \text{ J kg}^{-1}.$$

This is the amount of energy per unit mass that must be given to the parcel for the above process to take place.

Note that using the hydrostatic approximation, equation (8.16) can be expressed as

$$W = -R_d m \int_i^f (T'_{\text{virt}} - T_{\text{virt}}) d \ln p. \qquad (8.17)$$

In a $(T - \ln p)$ diagram the above equation is proportional to the area enclosed between the vertical profile of the environment and the vertical profile of the parcel's virtual temperature. Equation (8.16) or (8.17) gives the *maximum* work done by the buoyancy and when divided by the total mass defines (remember we are always dealing with adiabatic processes, i.e. $\delta Q = 0$) what is called the convective available potential energy (CAPE) and the convective inhibition energy (CINE). More on CAPE and CINE and their applications will be discussed in the next chapter.

Problems

(8.1) Elaborate on the five assumptions made in derivation of the stability conditions of a parcel that is subject to a small displacement from equilibrium.

(8.2) If the virtual temperature profile in a layer is $T_{virt}(z) = T_{\text{virt},0} a/(a + z)$, where $a > 0$ and $T_{\text{virt},0} > a\Gamma_d$, derive the vertical profile of the virtual potential temperature. What is the condition for stability for unsaturated parcels in this layer? (Unstable for $z < z_c$, stable for $z > z_c$ and neutral for $z = z_c$, where $z_c = \sqrt{(aT_{\text{virt},0}/\Gamma_d)} - a$)

(8.3) If the profile of virtual potential temperature in a layer is

$$\frac{\theta_{\text{virt}}}{\theta_{\text{virt},0}} = e^{\frac{z}{a_1}} - \frac{z}{a_2},$$

what is the condition for stability in the layer for unsaturated parcels? (Stable when $z > a_1 \ln(a_1/a_2)$, unstable when $z < a_1 \ln(a_1/a_2)$)

(8.4) An atmospheric layer is isothermal. A dry parcel is subjected to an upward displacement and begins to oscillate about its original level. Plot the oscillation period as a function of the temperature of the layer. What do you observe?

(8.5) If we define the geopotential ϕ according to $d\phi = gdz$ with $\phi(0) = 0$ show that under hydrostatic conditions

$$\Delta z = \left(\frac{R_d}{g} \ln \frac{p_1}{p_2} \right) \overline{T}_{\text{virt}},$$

where Δz is the thickness of a layer bounded by p_1 and p_2 ($p_1 > p_2$) and $\overline{T}_{\text{virt}}$ is the mean virtual temperature of the layer. Using the above relationship show that if a layer is lifted *en masse* while the mean virtual temperature remains constant, then the upper level pressure change, dp_2, is related to the lower level change, dp_1, via the equation

$$\frac{dp_1}{p_1} = \frac{dp_2}{p_2}.$$

(8.6) Show that for a constant virtual temperature lapse rate atmosphere $(T_{\text{virt}}(z) = T_{\text{virt}} - \Gamma_{\text{virt}} z)$

$$p = p_0 \left[1 - \frac{\Gamma_{\text{virt}} z}{T_{\text{virt},0}} \right]^{g/R_d \Gamma_{\text{virt}}}.$$

(8.7) Show that for an atmosphere where $T_{\text{virt}}(z) = T_{\text{virt},0}$

$$p = p_0 e^{-gz/R_d T_{\text{virt},0}}$$

(8.8) From the following data determine the stability and convective stability of each layer

p(mbar)	$T_{\text{virt}}(°C)$	$T_{\text{dew}}(°C)$
1000	30.0	22.0
950	25.0	21.0
900	18.5	18.0
850	16.5	15.0
800	20.0	10.0
750	10.0	5.0
700	−5.0	−10.0
650	−10.0	−15.0
600	−20.0	−30.0

(8.9) In an unstable layer of air extending from the ground to the pressure level of 900 mbar, the virtual temperature decreases with height at a rate of $25\ ^\circ\mathrm{C}\ \mathrm{km}^{-1}$. A parcel of air at the surface is given an initial velocity of $1\ \mathrm{m\ s}^{-1}$. If the virtual temperature at the surface is $7\ ^\circ\mathrm{C}$, and assuming that the parcel is and remains dry, find its position and velocity after one minute. $(81\ \mathrm{m}, 2.1\ \mathrm{m\ s}^{-1})$

(8.10) Sea-breezes and lake-breezes are established when the air over the land becomes warmer than the air over the water. The warmer air over land rises thus destroying the pressure stratification and resulting in horizontal pressure gradients between land and water. The height of the breeze is the altitude where the horizontal pressure gradient vanishes. Assuming that the vertical virtual temperature profiles over land and water are isothermal and the same, show that this height is given by

$$h = \frac{R_\mathrm{d}\ln(p_\mathrm{w}/p_\mathrm{l})}{g\left(\frac{1}{T_\mathrm{virt,w}} - \frac{1}{T_\mathrm{virt,l}}\right)}$$

where the subscripts l and w denote the conditions over land and water, respectively.

(8.11) Consider problems 8.6, 8.7, and 8.10. How would the results change if these problems were stated in terms of the temperature profiles rather than virtual temperature profiles?

CHAPTER NINE

Thermodynamic diagrams

Chapters 1 to 6 laid down the fundamental physical and mathematical concepts pertaining to thermodynamics. While we always kept the discussion close to our atmosphere, it was not until Chapter 7 that, in a more applied mood, we presented how these concepts are applied to yield quantities useful to atmospheric processes. However, even if we understand all the mathematics, we still need an efficient way to present and visualize thermodynamic processes in the atmosphere. Thermodynamic diagrams can do that.

Since the purpose of a diagram is efficiently and clearly to display processes and estimate thermodynamic quantities, the following are very desirable in a thermodynamic diagram: (1) for every cyclic process the area should be proportional to work done or energy (area-equivalent transformations), (2) lines should be straight (easy to use), and (3) the angle between adiabats and isotherms should be as large as possible (easy to distinguish). Up to now in our examples we have made use of the (p, V) diagram. This diagram satisfies the first condition $(pda = dw)$, but the angle between isotherms and adiabats is not very large (figure 4.5(a)). Because of this, while it is used for illustration purposes, this diagram is not used in practice.

9.1 Conditions for area-equivalent transformations

When we are constructing a new diagram, in effect we go from the $x = a, y = -p$ domain to a new domain characterized by two new coordinates, say u and w. These two coordinate systems are shown in figure 9.1. If we want the first condition to be satisfied then an area $dA = |\overrightarrow{dx} \times \overrightarrow{dy}|$ in the x, y domain is mapped onto an area $dA' = |\overrightarrow{d\mathsf{u}} \times \overrightarrow{d\mathsf{w}}|$ in the u, w domain, such that $dA = \zeta dA'$ ($\times$ here is the cross product between two vectors and $\zeta > 0$ is a

Figure 9.1
Area-equivalent
transformations.

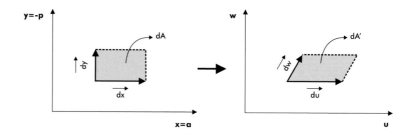

constant). Since any pair of points x and y are transformed to **u** and **w** we have that $x = x(\mathsf{u}, \mathsf{w})$ and $y = y(\mathsf{u}, \mathsf{w})$. We can thus write

$$\overrightarrow{dx} = \frac{\partial x}{\partial \mathsf{u}} \, \overrightarrow{d\mathsf{u}} + \frac{\partial x}{\partial \mathsf{w}} \, \overrightarrow{d\mathsf{w}}$$

$$\overrightarrow{dy} = \frac{\partial y}{\partial \mathsf{u}} \, \overrightarrow{d\mathsf{u}} + \frac{\partial y}{\partial \mathsf{w}} \, \overrightarrow{d\mathsf{w}} \ .$$

Then dA is equal to

$$dA = \left(\frac{\partial x}{\partial \mathsf{u}} \frac{\partial y}{\partial \mathsf{w}} - \frac{\partial x}{\partial \mathsf{w}} \frac{\partial y}{\partial \mathsf{u}} \right) \left| \overrightarrow{d\mathsf{u}} \times \overrightarrow{d\mathsf{w}} \right|, \tag{9.1}$$

where we have used the facts that

$$\overrightarrow{d\mathsf{u}} \times \overrightarrow{d\mathsf{u}} = \overrightarrow{d\mathsf{w}} \times \overrightarrow{d\mathsf{w}} = 0$$

and

$$\overrightarrow{d\mathsf{u}} \times \overrightarrow{d\mathsf{w}} = - \overrightarrow{d\mathsf{w}} \times \overrightarrow{d\mathsf{u}} \ .$$

In equation (9.1) $\left| \overrightarrow{d\mathsf{u}} \times \overrightarrow{d\mathsf{w}} \right| = dA'$. Thus,

$$dA = \zeta dA' \tag{9.2}$$

where

$$\zeta = J = \begin{vmatrix} \dfrac{\partial x}{\partial \mathsf{u}} & \dfrac{\partial x}{\partial \mathsf{w}} \\[2mm] \dfrac{\partial y}{\partial \mathsf{u}} & \dfrac{\partial y}{\partial \mathsf{w}} \end{vmatrix}$$

is the Jacobian of the coordinate transformation. It follows that the condition for area-equivalence is that the Jacobian should be a constant (if $J = 1$, then this is an equal-area transformation). The above approach tells us whether or not the two new coordinates provide an area-equivalent transformation, but both **u** and **w** must be specified. If only one is specified, then the following approach provides the other one. Since $dA = \zeta dA'$ it follows that

$$\oint y dx = \oint -p da = \zeta \oint \mathsf{w} d\mathsf{u}$$

or

$$\oint (pda + \zeta wdu) = 0.$$

The above condition is true if $pda + \zeta wdu = dz$ is an exact differential. Thus, $z = f(a, u)$ and

$$dz(a, u) = \left(\frac{\partial z}{\partial a}\right)_u da + \left(\frac{\partial z}{\partial u}\right)_a du.$$

It follows that the sufficient conditions for an area-equivalent transformation are

$$p = \left(\frac{\partial z}{\partial a}\right)_u$$

and

$$\zeta w = \left(\frac{\partial z}{\partial u}\right)_a.$$

From the above two equations we have

$$\left(\frac{\partial p}{\partial u}\right)_a = \frac{\partial^2 z}{\partial a \partial u}$$

and

$$\zeta \left(\frac{\partial w}{\partial a}\right)_u = \frac{\partial^2 z}{\partial a \partial u}.$$

Thus, if

$$\zeta \left(\frac{\partial w}{\partial a}\right)_u = \left(\frac{\partial p}{\partial u}\right)_a \qquad (9.3)$$

then the areas will be equivalent (again here if $\zeta = 1$ the areas will be equal). From equation (9.3) if we specify u we can determine what w should be in order for the transformation to be an area-equivalent transformation.

9.2 Examples of thermodynamic diagrams

9.2.1 The tephigram

If we suppose that $u = T$ then

$$\left(\frac{\partial p}{\partial u}\right)_a = \left(\frac{\partial p}{\partial T}\right)_a.$$

Using the ideal gas law, the above leads to

$$\left(\frac{\partial p}{\partial u}\right)_a = \frac{R}{a}.$$

From equation (9.3) we then have

$$\zeta \left(\frac{\partial \mathsf{w}}{\partial a} \right)_T = \frac{R}{a}$$

or

$$\zeta \left(\frac{\partial \mathsf{w}}{\partial a} \right)_T da = \frac{R}{a} da$$

or after integrating

$$\zeta \mathsf{w} = R \ln a + f(T) \qquad (9.4)$$

where $f(T)$ is the constant of integration which we can choose. By manipulating Poisson's equation we can write

$$\frac{T}{\theta} = \left(\frac{RT}{1000a} \right)^k$$

or

$$\ln a = \frac{1}{k} [\ln \theta - \ln T] + \ln T + \ln R - \ln 1000$$

or

$$R \ln a = c_p \ln \theta + f'(T).$$

If we combine the above equation with equation (9.4) and choose $f(T) = -f'(T)$ and $\zeta = c_p$, we arrive at an area-equivalent diagram with coordinates

$$\mathsf{w} = \ln \theta \qquad (9.5)$$
$$\mathsf{u} = T.$$

Obviously, in this diagram the dry adiabats are straight lines and perpendicular to the isotherms (figure 9.2). The equation for the isobars can be derived from Poisson's equation for $p = $ constant:

$$\ln \theta = \ln T + \text{constant.}$$

In a coordinate system $(\ln \theta, T)$ the above equation describes logarithmic curves. As expected, the work done on a unit mass of air in a reversible cycle $(-q)$ is proportional to the cycle: $-q = -\oint T ds = -c_p \oint T d \ln \theta = -c_p A_{\text{teph}}$ where A_{teph} is the area of the cycle in the tephigram[†] (positive if described counterclockwise). If the air is dry, $c_p = c_{pd}$.

[†] The word "tephigram" originated from the coordinates of this diagram, which are T and entropy. Entropy was originally denoted as ϕ (phi), but is now denoted by S.

Figure 9.2
Isobars, isotherms, and
dry adiabats on a
tephigram. Also shown is
the $w_s = 10$ g kg^{-1}
saturation mixing ratio
line (solid bold line) and
the pseudo-adiabat
$\theta_e = 40\,^\circ$C (broken line).

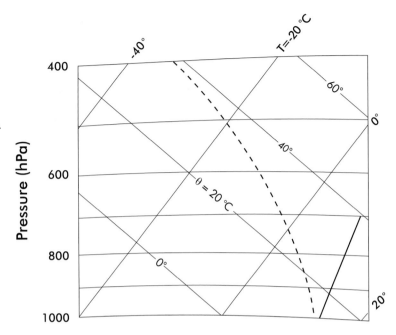

9.2.2 The emagram

Let us assume again that $u = T$. By taking the logarithm of the equation of state we get

$$\ln a = -\ln p + \ln R + \ln T.$$

Combining this equation with equation (9.4) yields

$$w = -\zeta^{-1} R \ln p + \zeta^{-1}[R \ln R + R \ln T + f(T)].$$

If we now choose $f(T) = -R \ln R - R \ln T$ and $\zeta = R$, we arrive at the area-equivalent diagram with coordinates

$$w = -\ln p \qquad\qquad (9.6)$$
$$u = T.$$

In this diagram (which is called an emagram – Energy per unit MAss diaGRAM) the isobars are parallel straight lines and perpendicular to the isotherms. The dry adiabats are again derived from Poisson's equation

$$\ln \theta = \ln T + k_d \ln 1000 - k_d \ln p$$

or

$$-\ln p = -\frac{1}{k_d} \ln T + \text{constant}.$$

Since the coordinates are $\ln p$ and T, the above equation is not a straight line. In the emagram the dry adiabats are slightly curved

Figure 9.3
Same as figure 9.2 but for
an emagram.

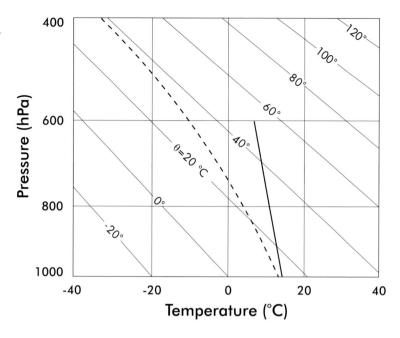

lines (figure 9.3). Their angle with the isotherms is about $45°$. Here, again, the work done on a unit mass of air in a reversible cycle $(-\oint pda)$ is proportional to the area of the cycle:

$$-\oint pda = \oint [-d(pa) + adp]$$

$$= \oint [-RdT + adp]$$

$$= -\oint RdT + \oint adp$$

$$= \oint adp$$

$$= \oint RT\frac{dp}{p}$$

$$= -R\oint Td(-\ln p)$$

$$= -RA_{em}$$

where A_{em} is the area in the emagram enclosed by the cycle, which is positive if described counterclockwise. If the air is dry, $R = R_d$.

9.2.3 The skew emagram (skew $(T\text{--}\ln p)$ diagram)

Let us now assume that $u = -\ln p$. Then

$$\zeta \left(\frac{\partial w}{\partial a}\right)_{\ln p} = -\left(\frac{\partial p}{\partial \ln p}\right)_a$$

or

$$\zeta \left(\frac{\partial \mathsf{w}}{\partial a} \right)_{\ln p} = -p$$

or

$$\zeta \left(\frac{\partial \mathsf{w}}{\partial a} \right)_{\ln p} da = -pda.$$

Integrating for $p = $ constant yields

$$\zeta \mathsf{w} = -pa + F(\ln p)$$

or

$$\zeta \mathsf{w} = -RT + F(\ln p).$$

We may choose $F(\ln p) = -\xi \ln p$ where ξ is a constant. Then

$$\zeta \mathsf{w} = -RT - \xi \ln p.$$

If we choose $\zeta = R$, we arrive at a new diagram with coordinates

$$\mathsf{w} = -T - \xi/R \ln p$$
$$\mathsf{u} = -\ln p.$$

Since the sign of an area in a diagram involves only the direction in which a cycle is carried out, the new coordinate system can be written as

$$\mathsf{w} = T + \frac{\xi}{R} \ln p \qquad (9.7)$$
$$\mathsf{u} = -\ln p.$$

This represents a correction (in fact a rotation) to the emagram which provides larger angles between any adiabats and isotherms (figure 9.4). An isotherm in this diagram has the equation

$$\mathsf{w} = \text{constant} + \frac{\xi}{R} \ln p$$

or

$$\mathsf{w} = \text{constant} - \frac{\xi}{R} \mathsf{u}.$$

In the (u, w) domain this equation represents a straight line with a slope equal to ξ/R. Thus, we may choose a ξ for which the isotherms are perpendicular to the dry adiabats. Note that the equation for dry adiabats is not linear. From Poisson's equation we have that $\ln \theta = \ln T + k_d \ln 1000 - k_d \ln p$ or $\ln p = (1/k_d) \ln T +$ constant. While here $\ln p$ is a coordinate, $\ln T$ is not and so the dry adiabats are not exactly straight lines. Here again it is easy to show that the work done in a cyclic process is proportional to the area of the cycle. The skew emagram and tephigram are the most commonly used diagrams. Other diagrams have been proposed (see problems 9.1–9.3), but their practical use is limited. A detailed

Figure 9.4
Same as figure 9.2 but for
a skew $(T–\ln p)$ diagram.

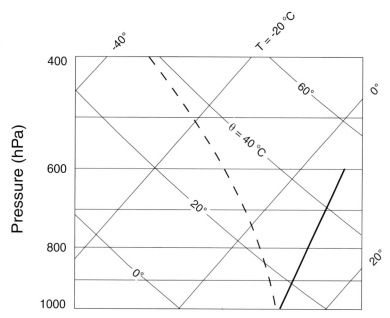

skew T–$\ln p$ diagram is supplied at the end of the book (figure A.1).

For a complete description of a thermodynamic process in a diagram, in addition to isotherms, isobars, and adiabats, we also need to define the constant mixing ratio lines and the saturated adiabats. From the definition of mixing ratio we have

$$p = e_{\mathrm{sw}}(T) + \frac{\epsilon e_{\mathrm{sw}}(T)}{w_{\mathrm{sw}}}$$

or

$$p \approx \frac{\epsilon e_{\mathrm{sw}}(T)}{w_{\mathrm{sw}}}. \tag{9.8}$$

For a constant w_{sw} the above equation defines a family of curves, which in the $(T, \ln p)$ domain are approximately (but not quite) straight lines:

$$p \approx \frac{6.11\epsilon}{w_{\mathrm{sw}}} \exp\left(19.83 - \frac{5417}{T}\right)$$

or

$$\ln p \approx \ln\frac{6.11\epsilon}{w_{\mathrm{sw}}} + 19.83 - \frac{5417}{T}$$

or

$$\ln p \approx A + \frac{B}{T}$$

where $A = \ln(6.11\epsilon/w_{\mathrm{sw}}) + 19.83$ and $B = -5417$. The above equation for typical values of w_{sw} and T is nearly a straight line.

Figure 9.5
Graphical procedure to
obtain thermodynamic
variables, from a skew
$(T-\ln p)$ diagram, given
an initial state (T, p).

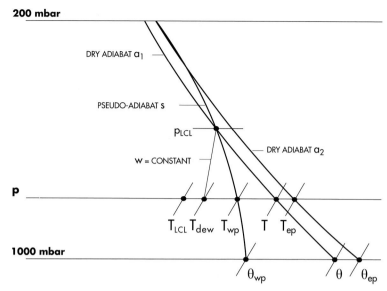

Since $w = w_{\mathrm{sw}}(T_{\mathrm{dew}}, p)$, the dew point of unsaturated air defines the mixing ratio w at (T, p). It follows, that the mixing ratio at (T, p) defines the dew point temperature.

The saturated adiabats are determined numerically from equation (7.63). They can be labeled in terms of the estimated θ_{ep} or in terms of the related (through equation (7.64)) θ_{wp}.

9.3 Graphical representation of thermodynamic variables in a diagram

Figure 9.5 shows the graphical procedure to obtain all the thermodynamic variables derived analytically in Chapter 7, for a skew $T-\ln p$ diagram. We start with the state (T, p) where the air is assumed to be unsaturated and to have a mixing ratio w. The intersection between the $w = $ constant line and the temperature axis gives the dew point temperature. By following the dry adiabat passing through point (T, p) (line a_1) down to 1000 mbar we obtain the potential temperature θ. By following it up until we intersect the $w = $ constant line we arrive at the lifting condensation level (p_{LCL}), where the parcel first becomes saturated and has a temperature T_{LCL}. Through point $(T_{\mathrm{LCL}}, p_{\mathrm{LCL}})$ passes a pseudo-adiabat s defined by θ_{ep}. If we follow this pseudo-adiabat down to the original level p, we obtain the pseudo-wet-bulb temperature, T_{wp}, and continuing down to the 1000 mbar level we obtain the pseudo-wet-bulb potential temperature, θ_{wp} (this also indicates the one-to-one relationship between θ_{wp} and θ_{ep} hence the labeling

Figure 9.6
Sounding at Davenport, IA on 21 June 1997 at 0000 UTC. As explained in the text, the information provided by this diagram would not support the development of deep convection that day. Nevertheless, deep convection did occur later in the day and very severe weather resulted in parts of the Midwest. See text for details.

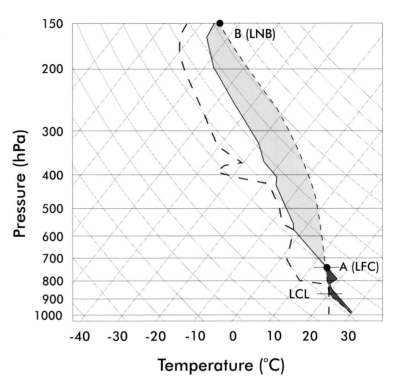

of pseudo-adiabats in terms of either θ_{ep} or θ_{wp}). If we follow the pseudo-adiabat s up to a pressure level where we can assume that all vapor has condensed and has fallen out of the parcel (typically this level is taken to be the 200 mbar level), and then follow the dry adiabat passing through that point (line a_2) down to the original level we obtain the pseudo-equivalent temperature, T_{ep}. Continuing down to the 1000 mbar level yields the pseudo-equivalent potential temperature, θ_{ep}. Note that T_{ei}, T_w, θ_w and θ_e are not directly estimated in diagrams. As we discussed in Chapter 7, even though they are very close to each other, in actuality $T_{ei} \neq T_{ep}$, $T_w \neq T_{wp}$, $\theta_w \neq \theta_{wp}$, and $\theta_{ei} \neq \theta_{ep}$ (recall equation (7.66)).

9.3.1 Using diagrams in forecasting

Figure 9.6 shows in a skew T–$\ln p$ diagram a sounding taken at Davenport, IA on 21 June 1997 at 0000 UTC. The solid line is the vertical profile of environmental temperature and the broken line to its left is the vertical profile of dew point temperature. The parcels at the surface are unsaturated. The lifting condensation level of these parcels is about 860 mbar where $T_{LCL} = 20$ °C. Above the LCL the dotted line is the pseudo-adiabat passing through the point $(T_{\mathrm{LCL}}, p_{\mathrm{LCL}})$ and below the LCL the dotted line is the dry adiabat passing through point (T, p) at the surface. The pseudo-adiabat intersects the temperature profile at point A. Up to this point the parcel is colder than the environment. As such the layer from surface to point A is stable. This means that the buoyancy in this layer is negative and that work must be done against the buoyancy if parcels originating at the surface are to reach that level. The work needed is proportional to the darker shaded area (negative area). This work or the magnitude of this negative area is often called the convective inhibition energy (CINE) given by equation (8.17) for $i = p_{\mathrm{surface}}$ and $f = p_A$ (here and in practice we make no distinction between T and T_{virt}).

$$\mathrm{CINE} = R \int_{p_{\mathrm{surface}}}^{p_A} (T' - T) d\ln p.$$

In our example, $p_{\mathrm{surface}} \approx 980$ mbar and $p_A \approx 740$ mbar. Assuming that the average $T' - T$ in this layer is about $-0.8\ ^{\circ}\mathrm{C}$ and that $R \approx R_{\mathrm{d}}$, we find that $\mathrm{CINE} \approx 64\ \mathrm{J\ kg}^{-1}$. If we now recall equation (8.15) we have that

$$\mathrm{CINE} = \int_{p_{\mathrm{surface}}}^{p_A} \frac{dv}{dt} v dt = \frac{1}{2} v_A^2 - \frac{1}{2} v_{\mathrm{surface}}^2$$

where v is the vertical velocity. Assuming that $v_A \approx 0$ it follows that if a parcel that originated at the surface were to reach point A, the *minimum* initial impulse has to be

$$v_{\mathrm{min}} = \sqrt{2\ \mathrm{CINE}} \approx 11.4\ \mathrm{m\ s}^{-1}.$$

Therefore, unless an initial impulse of $11.4\ \mathrm{m\ s}^{-1}$ is applied to the parcels they will not make it to point A. This point is called the level of free convection (LFC). Above this level the parcel is warmer than the environment all the way up to 150 mbar. The buoyancy from LFC to 150 mbar is positive and now work is done by buoyancy. This work is proportional to the lighter shaded area (positive area). In this layer the parcels that made it to LFC are now free to rise and accelerate. Note that owing to the temperature inversion in the stratosphere, above 150 mbar the temperature normally increases with height. As such, there is a point (B in our example) where the temperatures of the parcel and the environment become equal again. Above this level, which is referred to as the level of neutral buoyancy (LNB), work is done against buoyancy and the parcels begin to decelerate. The magnitude of the positive area is called the convective available potential energy (CAPE) and is given by

$$\mathrm{CAPE} = -R \int_{\mathrm{LFC}}^{\mathrm{LNB}} (T' - T) d\ln p.$$

Assuming that between LFC and LNB the average difference between the temperature of the parcels and the environment is about $7\,^{\circ}C$ and that $R \approx R_{\rm d}$ we can estimate that CAPE ≈ 3200 J kg^{-1}. As we did above, we can show that in this case the *maximum* velocity a parcel will attain is

$$v_{\rm max} = \sqrt{2\,{\rm CAPE}} \approx 80 \text{ m s}^{-1}.$$

Both CINE and CAPE are very useful as they provide information on whether or not convection will occur (CINE) and on how severe a storm might become (CAPE).

In our example CAPE is rather high. Thus, if the parcels make it to LFC, they will then accelerate greatly and deep convection will ensue. The problem in predicting deep convection in this particular example, however, is that the value of CINE is considerable. Such values may inhibit development of convection. In practice, situations of high CAPE values should be monitored to see whether the negative area is removed by warming of the air in the lower levels, or whether upward acceleration is aided by a low level jet or frontogenesis. In our case all of the above happened thereby allowing deep convection and severe weather to develop later in the day in the Midwest.

Together with CAPE and CINE we may calculate the precipitable water from soundings. If we recall the definition of precipitable water in problem 7.19 and the definition of specific humidity in equation (7.2) we have

$$d_{\rm w} = \frac{1}{\rho_{\rm w}} \int_0^\infty {\rm q}\rho dz$$

or using the hydrostatic approximation

$$d_{\rm w} = \frac{1}{g\rho_{\rm w}} \int_0^{p_{\rm surface}} {\rm q} dp.$$

The above equation indicates that precipitable water can be estimated by integrating the specific humidity profile. The specific humidity profile as a function of pressure can be obtained from the vertical profiles of T and $T_{\rm d}$ using the equations presented in Chapter 7.

Examples

(9.1) An air parcel at 1000 mbar has a temperature of 20 °C and a mixing ratio of 10 g kg^{-1}. The parcel is lifted to 700 mbar by passing over a mountain. During the ascent 80% of the water vapor that condenses falls out of the parcel. Find the temperature and potential temperature of the air when it returns to 1000 mbar on the other side of the mountain.

The parcel is located at the point with coordinates $T = 20\ °C$ and $p = 1000$ mbar ($w = 10$ g kg^{-1}). At this point, the saturation mixing ratio is ~ 16 g kg^{-1}. Therefore, the parcel is unsaturated and as it begins to rise it will follow the dry adiabat at that point (which because we are at 1000 mbar is $20\ °C$). The point at which this dry adiabat intersects the $w = 10$ g kg^{-1} line is at $p = 910$ mbar and $T = T_{\rm LCL} \approx 12.4\ °C$. This is the lifting condensation level and from that point on the parcel will follow the corresponding pseudo-adiabat ($\theta_{\rm ep} = 49.5\ °C$, which in the skew T–ln p diagram is labeled in terms of the related $\theta_{\rm wp} = 16.5\ °C$). At 700 mbar the saturation mixing ratio is ~ 6.66 g kg^{-1}. Therefore, $10 - 6.66 = 3.44$ grams of vapor per kilogram of dry air have condensed during the saturated ascent from 910 mbar to 700 mbar. Since 80% of this amount is removed by precipitation, it follows that 0.688 g kg^{-1} of liquid water remains in the parcel as it begins its descent. This means that initially the descent will be dictated by the same saturated adiabat until all the water evaporates, i.e. until the mixing ratio increases from 6.66 to $6.66 + 0.688 = 7.348$ g kg^{-1}. This will happen at the point where the pseudo-adiabat intersects the $w = 7.348$ line. This is the point with coordinates $p \approx 750$ mbar and $T \approx 4\ °C$. From this point on, the descent is dictated by the corresponding dry adiabat ($\theta \approx 28\ °C$). Thus, on the other side of the mountain at $p = 1000$ mbar, $T = \theta = 28\ °C$.

(9.2) Using a tephigram, demonstrate the criteria for convective instability stated in equation (8.14).

Let us consider an isothermal (and thus stable) layer extending from 1000 mbar to 950 mbar. Let us further assume that initially the layer is unsaturated and that it is lifted *en masse* until it becomes saturated in its entirety (a lifting of 300 mbar would normally suffice to saturate the layer). We can consider the following three possibilities for the layer to reach saturation. (1) All points in the layer after a preliminary dry adiabatic expansion reach saturation along the same pseudo-adiabat; (2) the top of the layer reaches saturation along a higher pseudo-adiabat than the bottom; (3) the top of the layer reaches saturation along a lower pseudo-adiabat than the bottom.

These three possibilities are presented in figure 9.7 where the initial and final positions of the layer are shown. In figure 9.7(a), in the final state every point in the layer is saturated and has the same pseudo-equivalent potential tem-

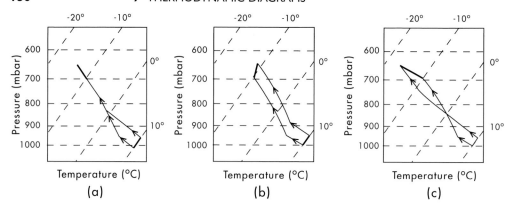

Figure 9.7 Illustration of criteria for convective instability discussed in example 9.2.

perature, θ_{ep}. Consequently, the lapse rate in the layer is exactly the saturated adiabatic rate. Thus, the layer has gone from stable to neutral. It follows that for an initial profile $d\theta_{ep}/dz = 0$, the saturated layer will be neutral with respect to saturated parcel processes. In figure 9.7(b) we see that, because the initial θ_{ep} profile in the layer corresponds to $d\theta_{ep}/dz > 0$, the final temperature profile in the layer connects two pseudo-adiabats with the bottom of the layer at the lower θ_{ep} and the top of the layer at the higher θ_{ep}. Thus, the final temperature profile corresponds to a temperature profile that is smaller than Γ_s. This means that the saturated layer has become even more stable with respect to saturated parcel processes. Figure 9.7c shows the third possibility. Here, the initial θ_{ep} profile in the layer corresponds to $d\theta_{ep}/dz < 0$, which results in a final temperature profile in the layer that decreases with height and it is steeper than the pseudo-adiabats. As such, the saturated layer has become unstable with respect to saturated parcels.

The above results are independent of the initial temperature profile in the layer. As such, these criteria are general.

Problems

(9.1) Show that the Refsdal diagram with coordinates $u = \ln T$, $w = -T \ln p$ is an area-equivalent diagram and that neither its isobars nor its dry adiabats are straight lines.

(9.2) Show that the Stüve (or pseudo-adiabatic) diagram with coordinates $u = T, w = -p^{k_d}$ is not an area-equivalent diagram. Show that isobars, isotherms, and dry adiabats are all straight lines.

(9.3) Using the Jacobian method show that the diagram with coordinates $u = T, w = c_p \ln \theta$ is an equal-area diagram.

(9.4) Given the sounding shown in the table,

p(mbar)	$T(°C)$
950	22.5
900	18.0
850	15.0
800	16.0
750	12.0
700	7.0
650	4.0
600	−1.5
500	−10.0
400	−20.0

determine using a diagram (1) the mixing ratio, (2) the relative humidity, (3) the potential temperature, (4) the pseudo-wet-bulb temperature, (5) the pseudo-wet-bulb potential temperature. Having determined the above, compute the wet-bulb temperature, the equivalent temperature, the pseudo-equivalent potential temperature, and the pseudo-equivalent temperature. Assume that $l_v = 2.45 \times 10^6$ J kg^{-1}, $c_w = 4218$ J kg^{-1}, and that the dew point at 950 mbar is 15.7 °C. (12.4 g kg^{-1}, 0.68, 27.5 °C, 18 °C, 20 °C, 19°, 51 °C, 66 °C, 60 °C)

(9.5) Approximate CAPE if in the previous problem we extend the sounding as follows:

p(mbar)	$T(°C)$
300	−35.0
200	−55.0
150	−60.0

Would you expect deep convection to occur?

(9.6) Consider the following scenario. On a given day, on a skew $T - \ln p$ diagram the virtual temperature profile from the surface (1000 mbar) to 200 mbar has a constant lapse rate of 7 °C km^{-1}. Above 200 mbar the temperature is constant. If the parcels at the surface are saturated, what will the maximum vertical velocity be on that day assuming that the saturated adiabatic ascent can be approximated by the constant lapse rate of 6.5 °C km^{-1}? The virtual temperature at the surface is 30 °C. Hint: use the results in problem 8.6 and a little bit of geometry. (~ 57 m s^{-1})

(9.7) A parcel of air with an initial temperature of 15 °C and dew point 0 °C is lifted adiabatically from 1000 mbar. Determine the lifting condensation level and temperature at that level. If the parcel is lifted a further 250 mbar what will its final temperature be and how much liquid water is produced? (~ 800 mbar, ≈ -3 °C, ~ -22 °C, ~ 2.7 g kg^{-1})

(9.8) In the previous problem if all the condensed water falls as precipitation, can you roughly estimate the height of water collected on a rain collector of area 1 m^2? Here you have to make your own assumptions. How does this result compare with the result obtained using the formula in problem 7.19? Can you explain any differences?

(9.9) An air parcel at 900 mbar has a temperature of 10 °C and a mixing ratio of 5 g kg^{-1}. The parcel is then lifted to 700 mbar by passing a mountain. If in this process 70% of the condensed water is removed by precipitation, what will its temperature, potential temperature, and pseudo-wet-bulb potential temperature be when it returns to its initial level of 900 mbar on the other side of the mountain? (~ 14 °C, $\sim$ 23.5 °C, ~ 11 °C)

(9.10) If a temperature profile in °C could be described as $T(z) = 28 - 8z$ for $z \geq 1$ km and $T = 20$ °C for $z < 1$ km, and the mixing ratio at the surface is 16 g kg^{-1}, estimate using a diagram the level of free convection. (~ 2.4 km)

CHAPTER TEN

Beyond this book

The previous chapters have presented the basics in atmospheric thermodynamics. As we know, in atmospheric sciences the ultimate goal is to predict as accurately as possible the changes in weather and climate. Thermodynamic processes are crucial in predicting changes in weather patterns. For example, during cloud and precipitation formation vast amounts of heat are exchanged with the environment that affect the atmosphere at many different spatial scales. In this final chapter we will present the basic concepts behind predicting weather changes. This chapter is not meant to treat the issue thoroughly, but only to offer a glimpse of what comes next.

10.1 Basic predictive equations in the atmosphere

In the Newtonian framework the state of the system is described exactly by the position and velocity of all its constituents. In the thermodynamical framework the state is defined by the temperature, pressure, and density of all its constituents. In a dynamical system such as the climate system both frameworks apply. Accordingly, a starting point in describing such a system will be to seek a set of equations that combine both the mechanical motion and thermodynamical evolution of the system.

The fundamental equations that govern the motion and evolution of the atmosphere (and for that matter of the oceans and sea ice) are derived from the three basic conservation laws: the conservation of momentum, the conservation of mass, and the conservation of energy. For the atmosphere the equation of state relates temperature, density, and pressure. In summary, from the conservation of momentum we derive the following set of predictive equations of motion (Washington and Parkinson (1986))

$$\frac{du}{dt} - \left(f + u\frac{\tan\phi}{a}\right)v = -\frac{1}{a\cos\phi}\frac{1}{\rho}\frac{\partial p}{\partial\lambda} + F_\lambda \qquad (10.1)$$

$$\frac{dv}{dt} + \left(f + u\frac{\tan\phi}{a}\right)u = -\frac{1}{\rho a}\frac{\partial p}{\partial\phi} + F_\phi \qquad (10.2)$$

$$\frac{d\omega}{dt} = -\frac{1}{\rho}\frac{\partial p}{\partial z} - g + F_z \qquad (10.3)$$

with

$$\frac{d}{dt} = \frac{\partial}{\partial t} + \frac{u}{a\cos\phi}\frac{\partial}{\partial\lambda} + \frac{v}{a}\frac{\partial}{\partial\phi} + \omega\frac{\partial}{\partial z}$$

$$u = a\,\cos\phi\frac{d\lambda}{dt}$$

$$v = a\,\frac{d\phi}{dt}$$

$$\omega = \frac{dz}{dt}$$

where u, v, ω are the horizontal and vertical components of the motion, ϕ is the latitude, λ is the longitude, a is the radius of the earth, f is the Coriolis force, and F is the friction force. In the above equations the forces that drive the motion are the local pressure gradients, gravity, Coriolis, and friction.

Equation (10.3) can be approximated by the hydrostatic equation (by assuming that $d\omega/dt = 0$ and $F_z = 0$)

$$g = -\frac{1}{\rho}\frac{\partial p}{\partial z} \qquad (10.4)$$

which relates density to pressure.

From the law of conservation of mass we can derive the continuity equation

$$\frac{\partial\rho}{\partial t} = -\frac{1}{a\cos\phi}\left[\frac{\partial}{\partial\lambda}(\rho u) + \frac{\partial}{\partial\phi}(\rho v\cos\phi)\right] - \frac{\partial}{\partial z}(\rho\omega) \qquad (10.5)$$

which provides a predictive equation for density. The last two equations in this basic formulation come from thermodynamics. The first law of thermodynamics (which expresses the law of conservation of energy)

$$C_p\frac{dT}{dt} - \frac{1}{\rho}\frac{dp}{dt} = \frac{dQ}{dt}, \qquad (10.6)$$

where dQ/dt is the net heat gain, provides a predictive equation for temperature. The above equations (10.1, 10.2, 10.3 (or 10.4),

10.5, and 10.6) make a system of five equations with six unknowns $(u, v, \omega, p, \rho, T)$. The equation of state

$$p = \rho R T \tag{10.7}$$

provides the additional equation that connects pressure, density, and temperature, thus resulting in a system of six equations (called the primitive equations system), with six unknowns (assuming of course that dQ/dt, F_λ, and F_ϕ are constants and known. Note, however, that this system is not a closed one because dQ/dt, F_λ, and F_ϕ must be determined from the other variables. These terms are very important for climate simulations but for short-term weather prediction they are often ignored.

10.2 Moisture

The above system of predictive equations was derived without the inclusion of moisture. Even though a model can be derived without moisture in it, including moisture can dramatically improve modeling and prediction. In a manner analogous to the continuity of mass, the changes of moisture must be balanced by the moisture's sources and sinks. An equation of continuity for water vapor mixing ratio can be written as

$$\frac{dw}{dt} = \frac{1}{\rho} M + E \tag{10.8}$$

where M is the time rate of change of water vapor per unit volume due to condensation or freezing, E is the time rate of change of water vapor per unit mass due to evaporation from the surface and to horizontal and vertical diffusion of moisture occurring on scales below the resolution of the model, and w is the mixing ratio. Often the above equation is combined with equation (10.5) to obtain

$$\frac{\partial(\rho w)}{\partial t} + \frac{1}{a \cos \phi} \left[\frac{\partial}{\partial \lambda}(\rho w u) + \frac{\partial}{\partial \phi}(\rho w v \cos \phi) \right] + \frac{\partial}{\partial z}(\rho w w) = M + \rho E. \tag{10.9}$$

When water vapor changes to water (or ice) and vice versa, heat is added to or removed from the atmospheric system. In this case $dQ = d(l w_s / T)$.

The above general and simple equations constitute only the starting point for studying and investigating the fascinating dynamics of atmospheric motion. The equations have to be modified and adjusted in order to include the effect of many other factors (for example, radiative processes) and to be applied over a specific range of scales (mesoscale models, for example). Such studies are crucial in the effort to understand the climate and ecology of our planet.

REFERENCES

(1) Bohren, C.F. & Albrecht, B.A., *Atmospheric Thermodynamics* (Oxford University Press, New York, 1998).

(2) Bolton, D., The computation of equivalent potential temperature, *Mon. Wea. Rev.*, **108**, 1046–1053 (1980).

(3) Emanuel, K.A., *Atmospheric Convection* (Oxford University Press, New York, 1994).

(4) Fermi, E., *Thermodynamics* (Dover, New York, 1936).

(5) Iribarne, J.V. & Godson, W.L., *Atmospheric Thermodynamics* (D. Reidel, Dordrecht, 1973).

(6) Salby, M.L., *Atmospheric Physics* (Academic Press, San Diego, 1996).

(7) Washington, W.M. & Parkinson, C.L., *An Introduction to Three-Dimensional Climate Modeling* (University Science Books, Mill Valley, California, 1986).

APPENDIX

Table A.1. *Units*

Physical quantity	Unit	
	MKS system	egs system
length	m	cm
time	s	s
velocity	$m\,s^{-1}$	$cm\,s^{-1}$
acceleration	$m\,s^{-2}$	$cm\,s^{-2}$
mass	kg	g
density	$kg\,m^{-3}$	$g\,cm^{-3}$
force	$kg\,m\,s^{-2}$(newton, N)	$g\,cm\,s^{-2}$(dyne,dyn)
pressure	$kg\,m^{-1}s^{-2}$(pascal, Pa)	$g\,cm^{-1}s^{-2}$(microbar, µbar)
energy	$kg\,m^2\,s^{-2}$(joule, J)	$g\,cm^2\,s^{-2}$(erg)
specific energy	$m^2\,s^{-2}$($J\,kg^{-1}$)	$cm^2\,s^{-2}$($erg\,g^{-1}$)

Relations:
$1\,N = 10^5$ dyn
$1\,Pa = 10$ µbar
$1\,bar = 10^6$ µbar $= 10^5$ Pa
1 atmosphere $= 1013$ mbar $= 1013 \times 10^2$ Pa $= 1013$ hPa
$1\,J = 10^7$ erg
$1\,cal = 4.185\,J$

Table A.2. *Selected physical constants*

General constants

Avogadro's number, N_α	6.022×10^{23} mol^{-1}
Boltzmann's constant, k	1.38×10^{-23} J K^{-1}
Gravitational acceleration at sea level, g	9.807 m s^{-2}
Universal gas constant, R^*	8.314 J K^{-1} mol^{-1}

Constants for dry air

Molecular weight, M_d	28.97 g mol^{-1}
Density, ρ_d	1.293 kg m^{-3} (at STP)*
Specific gas constant, R_d	287 J kg^{-1}K^{-1}
Isobaric specific heat (at 273 K), c_{pd}	1005 J kg^{-1}K^{-1}
Isochoric specific heat (at 273 K), c_{Vd}	718 J kg^{-1}K^{-1}
Ratio of specific heats, γ_d	1.4

Constants for water

Molecular weight, M_v	18.015 g mol^{-1}
Density (water)	1000 kg m^{-3} (at STP)
Density (ice)	917 kg m^{-3} (at STP)
Specific gas constant, R_v	461.51 J kg^{-1}K^{-1}
$\epsilon = M_d/M_v = R_d/R_v$	0.622
Isobaric specific heat (at 273 K) (vapor, c_{pv})	1850 J kg^{-1}K^{-1}
Isochoric specific heat (at 273 K) (vapor, c_{Vv})	1390 J kg^{-1}K^{-1}
Specific heat capacity (at 273 K) (water, $c_{pw} \approx c_{Vw} \approx c_w$)	4218 J kg^{-1}K^{-1}
Specific heat capacity (at 273 K) (ice, $c_{pi} = c_{Vi} = c_i$)	2106 J kg^{-1}K^{-1}
Latent heats	see Table A3

STP: Standard temperature and pressure (273 K, 1013 mbar)

Table A.3. *Specific latent heats*

$T(°C)$	$l_s(10^6 \text{ J kg}^{-1})$	$l_f(10^6 \text{ J kg}^{-1})$	$l_v(10^6 \text{ J kg}^{-1})$
−100	2.8240		
−90	2.8280		
−80	2.8320		
−70	2.8340		
−60	2.8370		
−50	2.8383	0.2035	2.6348
−40	2.8387	0.2357	2.6030
−30	2.8387	0.2638	2.5749
−20	2.8383	0.2889	2.5494
−10	2.8366	0.3119	2.5247
0	2.8345	0.3337	2.5008
5			2.4891
10			2.4774
15			2.4656
20			2.4535
25			2.4418
30			2.4300
35			2.4183
40			2.4062
45			2.3945
50			2.3893

Figure A.1
The skew $T - \ln p$
diagram.

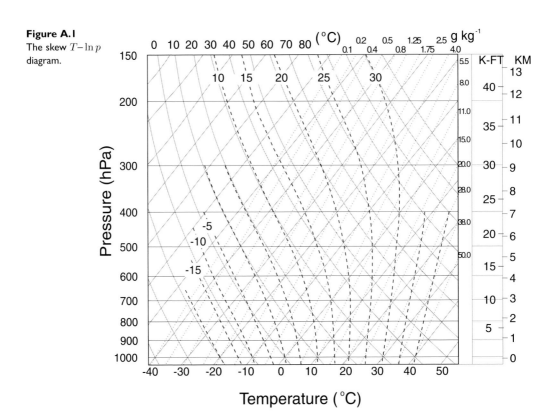

INDEX

absolute humidity, 94
absolute temperature, 13, 53–4
Amagat–Andrews diagram, 77
Avogadro
  hypothesis, 15
  number, 15

Boltzmann's constant, 8
Boyle's law 14, 15
Brunt–Väisälä frequency, 131
buoyancy, 130, 153, 154

Carnot cycle, 49–52
  efficiency, 52–4
  lessons learned from, 52–6
chemical potential, 73
  Chinook wind, 123
Clausius–Clapeyron equation, 79–81, 99, 105, 108, 115–16, 121
  and changes in the melting and boiling points of water, 85
  approximations of, 81–5
Clausius postulate, 53
cloud formation
  on ascent, 105
  on descent, 105
coefficient of thermal expansion
  isobaric, 11
  isochoric, 12
continuity equation, 160
convective available potential energy (CAPE), 139, 153–4
convective inhibition energy (CINE), 139, 153
convective stability, 136
critical point, 76

Dalton's law, 17–19, 93
degrees of freedom
  in Hamiltonian dynamics, 7
  in thermodynamics, 7
destabilization of lifted layers, 135–6
dew point temperature, 98–9, 100, 103
disorder, 62–3
dry adiabatic
  processes, 34–6
  lapse rate, 37–8
dry air
  composition, 19
  molecular weight, 19

energy
  conservation of, 4
  definition of, 25–9
  internal, 4
  kinetic, 4
  potential, 4
  relation between internal energy and temperature, 30
enthalpy
  definition of, 29
  of fusion, 78
  of sublimation, 78
  of vaporization, 78
  of temperature dependence of, 81–2
entropy
  and atmospheric processes, 66–7
  and disorder, 62–3
  and irreversible processes, 55–6
  and potential temperature, 64
  and reversible processes, 55
  definition of, 55

for ideal gases, 56
for liquids and solids, 56
maximization, 65
of an isolated system, 58
of a mixture, 67–8
equal distribution of energy theorem, 7
equations of motion in the atmosphere, 160
equivalence between heat and work, 27
equivalent (isobaric) potential temperature, 114
equivalent potential temperature, 111
equivalent (isobaric) temperature, 102–3
equilibrium
  between vapor and liquid, 74
  metastable, 3
  stable, 3
  unstable, 3
  vapor pressure
    for ice, 75
    for water, 75
    temperature dependence of phase transformation, 78
equilibrium state, 2
exact differential
  definition, 5
  conditions for, 6
fundamental relations, 60
fusion, 75
  latent heat of, 78

gas constant
  for dry air, 19
  for moist air, 96
  specific, 16
  universal, 16

Gibbs formula,   75
Gibbs function,   60
Guy-Lussac
    1st law, 11, 13
    2nd law, 12–13

heat capacities (*see* thermal
    capacities)
Helmholtz function,   60
hydrostatic approximation,
    37, 65, 115, 129
humidity
    absolute,   94
    relative,   94
    specific,   94

ideal gas
    definition,   8
    law,   10, 15–17

Joule's law,   30, 90

Kelvin's postulate,   52–3
kinetic theory of heat,   7

latent heat
    definition of,   78
    of fusion,   78
    of sublimation,   78
    of vaporization,   78
lifting condensation level (LCL)
    definition of,   106
    estimation,   107–8

Maxwell's relations,   70
mixing and irreversibility,   68
mixing ratio,   94
    saturation,   94
    water,   100
mixture of gases,   17–19
moist adiabatic
    lapse rate,   104
    processes,   100, 103, 104
moist air
    definition of,   93
    mean molecular weight,   95
moisture in the equations of
    motion,   161
molar heat capacity,   29
molecular weight of dry air,
    19

Poisson's equations,   34–6
potential temperature,   38–9,
    103
    temperature of unsaturated
        moist air,   103
process
    adiabatic expansion of
        unsaturated moist air,
        103–4
    adiabatic isobaric,   100

adiabatic isobaric mixing,
    116–17
isenthalpic,   100
isentropic,   109
isobaric cooling,   97–9
pseudo-adiabatic,   112-13
saturated adiabatic,
    109–112
saturated ascent,   109
vertical mixing,   118
pseudo-equivalent temperature,
    113–14
    equivalent potential
        temperature,   113
    wet-bulb temperature,   114
    wet-bulb potential
        temperature,   114

raindrops, growth by
    condensation,   87
relative humidity,   94

saturated adiabatic lapse rate,
    116
saturation temperature (at
    LCL),   106
specific heat capacity
    at constant pressure,   29
    at constant volume,   29
specific humidity,   94
specific gas constant,   16
stability criteria,
    for saturated parcels,   134–5
    for unsaturated parcels,
        133, 135
    of parcels when the layer is
        lifted,   135–6
stability, convective,   136
state
    equation of,   2, 10
    functions,   2
    mechanical,   1
    thermodynamical,   2
    variables,   2
sublimation,   75
    latent heat of,   78
supercooled water,   75, 88–9
supersaturation,   117–18
system
    closed,   1
    homogeneous,   1–2, 73
    heterogeneous,   2, 73
    isolated,   1
    open,   1

table of constants,   166
table of units,   165
temperature
    dew point,   97–100
    equivalent (isobaric), 102–3
    equivalent (isobaric)
        potential,   114

equivalent potential,   111
Kelvin scale,   13
potential,   38–9, 103
    of unsaturated moist air,
        103
pseudo-equivalent,   113–14
pseudo-equivalent potential,
    113
pseudo-wet-bulb,   114
pseudo-wet-bulb potential,
    114
relationship between dew and
    wet-bulb,   103
saturation (at LCL),   106
thermodynamic definition of,
    61
virtual,   96
virtual potential,   104
wet-bulb,   103
wet-bulb potential,   114
temperature lapse rate
    dry adiabatic,   37–8
    moist adiabatic,   104
    saturated adiabatic,   116
thermal (heat) capacity
    at constant pressure,   99
        for dry air,   32
        for moist air,   96–7
    at constant volume,   99
        for dry air,   32
        for moist air,   96–7
    of diatomic gases,   32
    of monatomic gases,   31
thermodynamics
    definition of,   1
    statistical nature of,   61–4
    1st law of,   27–8, 33
    2nd law of,   56, 58
    combining the two laws,   60
thermodynamic diagrams
    and forecasting,   153–4
    conditions for area-equivalent
        transformation,   143–5
    emagram,   147–8
    graphical representation in,
        151–2
    skew emagram (T–$\ln p$
        diagram),   148–50
    tephigram,   145–6
transformations
    adiabatic,   3, 34, 59
    cyclic,   3, 34
    irreversible,   3, 6
    isentropic,   59, 65
    isobaric,   3, 33, 59
    isochoric,   3, 33, 59
    isothermal,   3, 33, 59
    polytropic,   36–7
    reversible,   3, 6

units,   165
universal gas constant R^*,   16

vaporization, 75
  latent heat of, 78
variables
  dependent, 2
  extensive, 2
  independent, 2
  intensive, 2

virtual potential temperature,
  104
virtual temperature, 96

water
  density of, 76
  mixing ratio, 100
  precipitable, 127, 154

supercooled, 75, 88–9
  transformations of, 74–6
wet-bulb temperature, 103
wet-bulb potential temperature,
  114
work
  and kinetic energy, 25
  definition of, 23